THE BIBLE, NATURAL SCIENCE, AND EVOLUTION

Other Books of Dordt College Press

The Unity in Creation	Russell Maatman
The Armor of Light	Merle Meeter
Sign of a Promise	James C. Schaap
International Politics and Global Justice	James N. Skillen

THE BIBLE, NATURAL SCIENCE, AND EVOLUTION

RUSSELL W. MAATMAN

Dordt College

Dordt College Press
Sioux Center, Iowa 51250

Published by Dordt College Press
Sioux Center, Iowa 51250
Printed and bound in the United States of America, 1980

ISBN 0-932914-03-9

PREFACE

The Dordt College faculty has been generally dissatisfied with the traditional idea of separating the branches of learning from the Christian faith. Consequently, there is on the Dordt campus a continuing discussion of how the Christian faith and the academic disciplines ought to be integrated. One of the products of this discussion is *Scripturally-Oriented Higher Education,* the College's recently-adopted statement of purpose. The discussion has also stimulated faculty members to speak and to write. It would be difficult to find another campus which is as alive to meaningful issues as the campus of Dordt College. It is in this context that this book has been written.

I would like to acknowledge the particular help of four persons. Rev. B. J. Haan, Rev. J. B. Hulst, and Dr. Simon Kistemaker made helpful suggestions concerning the text. My wife, Jean, worked with me many hours, and I have learned that her patience is very great. Even though I have benefited from the insights of many others, none of them should be held responsible for what I have written. Anything that is in error is certainly my fault.

It is my prayer that this book will glorify the Triune God. If what I have written moves the reader to have a greater appreciation for the work of the Creator, my purpose will have been accomplished.

R. W. Maatman

Sioux Center, Iowa
October, 1969

CONTENTS

CONTENTS

1 Introduction

For more than a century there has been, in the minds of many Christians, a conflict between science and the Christian faith. Today the Christian asks, "What should be my attitude toward the modern explosion of scientific knowledge?"

Prior to the middle of the nineteenth century, before the effects of the explosion of scientific knowledge were felt, the individual Christian did not feel the necessity to develop an attitude toward science. Today he realizes that his attitude toward science cannot be neutral. Yet, before the middle of the nineteenth century, the Christian religion claimed, as it claims now, the whole devotion of its adherents. In the light of this claim the Christian wonders what the drastic change from a nonscientific to a scientific environment ultimately means. How does this change fit into the life of the individual Christian?

The Christian and the Natural Man

As questions related to the interaction between the Christian faith and science are discussed, it is possible to consider a wide variety of convictions. However, the approach of this book is based upon the claim that there are only two fundamental positions, that of the Christian and that of the natural man. These positions are described briefly now, and more fully in subsequent chapters.

The Christian assumes that the Bible is a revelation of God, whose existence is independent of the universe he created. The natural man assumes that the universe is just that which man can, in principle, comprehend, and he admits of the existence of nothing outside the universe. The universe is self-contained and there is no God.

The term "atheist" is not used in this book to designate the natural man because "atheist" is a negative term. The natural man is of course an atheist, but he is more than that. The term "natural man" indicates positively that which is central in this man's thinking, namely, nature. What God is to the Christian, nature is to the natural man. The natural man admits that man is part of nature. However, since man's ability, in principle, to observe and comprehend a thing is the test of whether or not that thing exists, man is in a sense put at the head of nature.

"Nature" as understood by the natural man is ultimately that which can be reduced to the physical. For the sake of convenience, this definition of "nature" is the one used in this book. There is, however, no intent to overemphasize the physical aspect of the universe. This deliberate effort to limit most of the discussion to the physical is made because of practical limitations. When, for example, man is discussed in this book there is no intention to indicate that man is the sum of his physical parts.

Since "nature" in the ensuing discussion denotes the physical universe, the terms "natural law" and "science" are also used in reference to the physical universe. In this discussion a natural law is a law which operates in this physical universe; such a law is the law of gravitation. By "science" is meant "natural science," i.e., the study of the physical universe and the laws operative within it. These are admittedly narrow definitions. But, while there is no quarrel with broader definitions, a clear understanding of what is meant by the above terms will be very useful in the discussion which follows.

By maintaining that there are only two fundamental positions, it is not meant that people are to be judged according to their conformity to one or the other of these positions, and then labelled "Christian" or "natural man." People are, of course, either Christian or non-Christian. Yet, in the thinking of a particular Christian there may be elements which are based on the assumptions of the natural man. Probably all Christians are in this category. Therefore, the "Christian position" and the "natural man's position" are idealized concepts. When in subsequent chapters a certain belief is condemned, it does not mean that the person having such a belief is likewise condemned. Men are not consistent. Those who are the most consistent natural-man thinkers are certainly guilty of grave error, but it is not the intent that such individuals be identified here, even though references to the literature affording examples of such thinking are occasionally given.

Is it not simplistic to suggest that there are only two fundamental positions among the many world-and-life views? There are, after all, countless non-Christians who believe in a god. If one holds that there is a god outside the universe who is not however the God of the Bible, is not this a belief which contradicts both the Christian position and the natural man's position?

Accepting a god who is not the God of the Bible seems at first to be a third possible position. Furthermore, it seems to correspond to the belief of most persons. A close examination of this supposed third position indicates that it is virtually equivalent to the position of the natural man. Both the natural man and the man who believes in a non-Christian god depend entirely on human authority. The natural man is consistent: he understands that he cannot "discover" a god outside the universe. The man who maintains that a non-Christian god exists makes such a claim, not because this god reveals himself to man, for there is no such god. Rather, he makes such a claim because his mind decides there is such a god.

Is it fair to assume that the non-Christian who believes in a god is wrong? It is appropriate to investigate whether it is possible, in the present discussion, to take a neutral position concerning the existence of a god, and, if a god does exist, which god it is that exists.

One cannot assume that all three of the postulated positions – those of the Christian, the non-Christian who believes in a god, and the natural man – can be simultaneously wrong. Either there is a god outside the universe, or there is not. Often it is attempted to base a discussion similar to the one in this book on the neutral position. The neutral position, even though it is an agnostic position, cannot provide the correct starting point because it provides no starting point. Any philosophical discussion, such as the one of interest here, involves an assumption, perhaps not stated explicitly, concerning the existence or non-existence of a god. For, if a discussant claims that he makes no assumption, he thereby indicates that his world-and-life view does not depend upon a god. But, building a universe without a god is precisely what the natural man does. The only difference between the hypothetical discussant and the natural man is that the natural man does not introduce an unnecessary, and therefore disposable, assumption into the structure of his universe. The neutral position is neutral in name only.

Thus it should be assumed that those who develop world-and-life views do make an assumption concerning whether or not a god outside this universe exists. Both the non-Christian who believes in a god and the Christian will agree that the natural man is wrong; and it can be seen that there is no position, besides these three, to consider. Either the Christian or the non-Christian who believes in a god is wrong. One of them has created a god, and his faith is actually only faith in his own imagination. The one who has created a god has, even though he does not realize it, made his own mind the judge of the universe. He may have constructed a system which does not look like the system of the natural man, but because that system is the product of his own mind he has used the same authority as the natural man, the mind of man. If it is remembered that the position being discussed is a *wrong* position, it seems extremely likely that those who adopt this wrong position will eventually consciously adopt the position of the natural man. Time will accomplish the merger of the two positions.

Thus, whoever is wrong, the Christian or the non-Christian who believes in a god, there remain only two basic positions: that of the natural man and that of the man who believes in the true God. In this book it is assumed that it is the Christian God who exists, and therefore the Christian position and the natural man's position are taken to be the only two basic positions. A non-Christian who believes in a god could write a book on the interaction between his faith and science. Such a book also ought to claim that there are but two basic positions. In other words, if a believer in a god claims for his god the very opposite of what the natural man claims, such a believer will *always* describe the interaction between his faith and science in terms of only two basic positions.

If there is a god, who is correct – the Christian, or the non-Christian who believes in a god?

Non-Christian religions do not claim that man is utterly helpless. The Christian religion is unique in its claim that man is completely unable to save himself. It is the Christian religion which teaches that all men offend God, and that no man can save either himself or any other man. Non-Christian religions teach or imply in some way that man has at least some sovereignty. There is nothing in any non-Christian religion which is like the Christian teaching concerning the grace of God. The concept of grace is so peculiarly Christian that it is difficult even to explain the concept to the non-Christian.

But, when one considers the contrast between the positions of the natural man and the believer in a god, it is precisely this dependence of man on a force outside the universe, this inability, in principle, of man to comprehend the universe, which distinguishes the two positions. In other words, if there is to be a meaningful alternative to the natural man's position, it is the Christian position which ought to be chosen. The hypothetical non-Christian religion which was mentioned as possible alternative to the position of the natural man does not correspond to an existing non-Christian religion. A careful analysis of assumptions indicates that it is the Christian who has something to say to the natural man. It is the Christian who has the alternative. Very likely the sharp distinction between the position of the one who believes in a god and the position of the natural man would not be made by anyone were there no Christian insistence on this matter. In parts of the world having little or no Christian influence, the principial differences between the religionist and the natural man concerning science, are small differences, compared to those areas which have at least some Christian cultural influence.

In spite of all that has been said above, the Christian-natural man difference upon which this book is based can be considered by the skeptical reader to be only a working hypothesis. If any reader can benefit from the following chapters, then it is possible that those who are not willing to make the assumption concerning the difference between the Christian and the natural man can also benefit. In any event, it ought to be evident that if a theistic position other than the Christian position is taken, then also there are only two basic positions to consider. Even then, the basic assumption made in this book can in many places be used profitably in the ensuing discussion.

The Pattern of the Book

What has been stated thus far includes reasons why the Christian position is preferable to a non-Christian theistic position, providing it is assumed that a god exists. It is necessary to be more specific about the Christian position. The Christian position should be Biblically-based in a very careful way. It is not *a* Christian position which is presented, but it is claimed that a Biblically-based Christian position is *the* Christian position, and that it is this position which is to be used when the relation between the Christian faith and science is discussed. This discussion begins in Chapter 2 with a consideration of part of the history of the interaction between the Christian faith and science. It is the historical relation between the Bible and science, and not the Christian faith and science, which is considered. Then, in Chapter 3, the perfection of the Bible is described and seen as the origin of the Christian position. The special relation between Biblical perfection and science is discussed.

In Chapters 4-7 other aspects of the relation between the Christian faith and science are examined. These chapters contain discussions of natural law, miracles, the natural man's view of the universe and the inherent limitations of this view, the dignity and importance of science (in contrast with a rather common, popularized debunking of science), and the God-created role of man in scientific endeavor.

To the extent the ideas in Chapters 3-7 are correct, they should be useful in analyses of areas of conflict between the Christian faith and science. Therefore, an analysis of one of those areas, the debate concerning creation and evolution, constitutes the remainder of the book. It is intended that the latter part of the book be first of all an example of how specific questions in the Christian faith-science interaction should be approached. Chapters 1-7 therefore serve as a background for the remainder of the book, although the first part of the book should not be considered to be *only* a background for such discussions. However, if the principles in the background material are correct, and if the evolution discussion given is a correct application of those principles, then one could use the same principles and procedure in working out other specific questions related to the Christian faith-science interaction.

The part of the book in which evolution is discussed is to serve not only as an example of the principles developed in the first part, but also as an attempted answer to the questions many people ask in this very modern, and very emotional, controversy. This controversy is not important merely because it interests and disturbs so many persons. It is also important because the questions raised are at the heart of any world-and-life view. When we concern ourselves with the interaction between the Christian faith and science, we touch upon that which is as important as anything in our lives. When we discuss evolution, we discuss as vital a specific problem concerning the work of God as we will ever encounter.

2 *Has the Bible Helped Science?*

Some Christians believe that a scientist who is a Christian can witness best to his non-Christian colleagues if he does not put the Bible in a pre-eminent position when the Christian faith is first presented. The Bible, it is said, is not accepted by the unbeliever and the Christian must be circumspect in introducing it to him. In this chapter there is no attempt to prove to the unbeliever that the Bible must be accepted. Instead, there is discussed a certain historical relation between the Bible and scientific activity, a relation not usually realized. The emphasis in this chapter is on a *historical* relation. In succeeding chapters the relation between science and the Bible is discussed.

The main thesis of this chapter is that modern science as we know it would not exist if it were not for a widespread dissemination of the Bible in certain places at certain times in history. In the first part of the discussion, an attempt is made to show that (1) wherever in modern times the Bible has been widely and seriously accepted, science has flourished and (2) this scientific growth has been necessary for modern science to exist. In the remainder of the discussion, the "why" of the connection between widespread dissemination of the Bible and science is discussed.

Before beginning the discussion, two phrases used must be defined. First, "growth of science" (or, "appearance of modern science") means in the context of this chapter the explosion of scientific knowledge which began in the last part of the eighteenth century and which continues up to now. "Growth of science" in this context is not the long, slow accumulation of scientific knowledge which took place throughout the world during the millennia prior to the present explosion. This slow accumulation was a necessary prerequisite for the modern explosion. For example, the ancient Greeks and others contributed many necessary mathematical ideas. Practical developments, such as the invention of gunpowder and of the compass by the Chinese, came from many places and times. In the centuries just prior to the beginning of the scientific knowledge explosion, there were men such as Newton, Galileo, and Copernicus, all of whom contributed vital concepts. There were also over the centuries the alchemists, for example, who contributed both correct and incorrect ideas concerning the nature of matter and of chemical reactions. The kind of stimulus the Bible provided for the modern explosion may or may not have been a factor in some of these earlier developments, but that period is prior to the period of the present discussion.

Second, the "widespread dissemination of the Bible," referred to frequently in what follows, means something far more specific than the formal acceptance of the Christian faith by large numbers of people, as for example, by whole populations. It cannot be assumed that merely because a state formally accepted the Christian faith, leading perhaps to the founding of a state church, that there was therefore a widespread dissemination of the Bible in that place. Perhaps some of the Bible-science interaction has been overlooked just because a careful distinction was not made between states in which there was a widespread dissemination of the Bible, and states in which there was not such a dissemination, even though there was a formal acceptance of the Christian faith.

Time and Place of the Explosion and of Bible Dissemination

By the end of the eighteenth century, when the modern explosion began, the scientific knowledge which had accumulated was available virtually everywhere. Who was it who used this accumulation so as to produce the modern scientific explosion? Which culture or cultures took advantage of the knowledge which had been gained? Who gave us modern science?

Contributions to scientific knowledge are *today* made by nearly every culture. Any criterion one could use – the number of scientific articles appearing each year, the number of persons engaged in scientific research, the amount of money spent on scientific research, or some other criterion – indicates that present-day science is not national or regional. One could show that Great Britain, Japan, the United States, Australia, India, Canada, Israel, Russia, South Africa, Argentina, Germany, China, Egypt, and many others are all making significant scientific contributions. Thus the question is not, "What is happening today?" but rather, "How did it all come about?"

The explosion of scientific knowledge began in northwestern Europe and North America. The productive countries were Germany, England, the United States, and several of their neighbors. In the earlier years of the period since the end of the eighteenth century, scientific activity in France was somewhat stifled, and a significant amount of the later French motivation seems to have come from outside influence. In Germany and England the emphasis was on the development of basic science; in the United States, on new inventions and engineering. These emphases spread to some neighboring countries, and the sum of this activity was unlike anything the world had yet seen.

In countries not in these productive regions – for example, Russia, Japan, and India – the present intense activity in science is built not so much upon their own nineteenth century achievements, as upon the achievements of the European and North American countries cited. There was, of course, some scientific activity in countries such as Russia during that period; but this level of activity had been maintained in very many places for centuries. In the productive countries the intellectual climate encouraged a breaking away from the old patterns. In the United States, for example, the contributions of an Edison were eagerly accepted, even though his origins were outside the intellectual class.

Russians claim that nineteenth-century Russia did make significant scientific contributions, citing the famous chemist, D. Mendeleev, who made key contributions to the all-important periodic classification of the elements. Yet, his career proves that life in nineteenth-century Russia was the very opposite of what is needed for science to grow. His widowed mother was able only through influence to enroll him in college; his graduate training was outside of Russia, and only through his efforts late in the nineteenth century did it become possible for others to receive graduate training in Russia. Mendeleev himself spoke out against the Russian government's oppression of students. With respect to the growth of scientific knowledge, nineteenth-century Russia seems to have been fundamentally the same as the Russia of earlier centuries.[1] With variations the same picture could be painted for the other areas outside northwestern Europe and North America.

The scientific activities of the productive countries since the end of the eighteenth century can be divided into two categories. First, there has been the development of almost all of the important ideas of the sciences. In physics, these ideas include our ideas about the structure of matter, the nature of radiation, the nature of radioactivity, quantum mechanics, electromagnetism, and relativity. In chemistry, there are included the development of the concept of the atom, the understanding of the combining tendencies of atoms in chemical reactions, the concepts of molecular structure, and many other important ideas. In biology, the important ideas developed include the application of some of these physical and chemical principles to living systems. During this period, all the basic ideas in geology were developed. Astronomy had been developing thousands of years; but almost all the major ideas of this science, particularly those which are applications of physical and chemical principles, have been conceived during the modern period.

Second, there were in the productive countries not only these basic scientific discoveries, but very many lesser scientific developments which have implemented the basic discoveries. Even before modern Russian, Japanese, etc. research made its impact, there were large numbers of people engaged in all the sciences in the productive countries. Often non-scientists do not realize that for the work of an Einstein or a Rutherford to make its impact, there must be many others who make lesser but necessary steps forward. There are today over the whole world more than one hundred thousand people engaged in scientific research, and who are therefore contributing to scientific progress. This necessary scientific activity on the part of such a large number of people *began* in the productive countries mentioned, the same countries in which the major breakthroughs were made.

It is attempted in what follows to show that the educational climate which is a prior necessity for an explosion of scientific knowledge was produced in these productive countries because in just those places large numbers of people received and knew the teachings of the Bible.

It was reasoned by those who spread the Christian message that the common man could not have the Bible in his heart unless he could read it. Therefore, public school systems were begun. In the productive countries Bible dissemination was usually synonymous with the founding of the public school system. The spread of Protestantism is often given as the motivation for the founding of public school systems:

> In the nearly three hundred years which separated the Reformation from the French Revolution considerable advance had been made, in Protestant countries, in abolishing illiteracy. Educational activity increased in Catholic countries also, though some of them, such as Italy and Spain, ran far behind their Protestant neighbors in providing instruction for the masses.[2]

The spread of Protestantism is essentially equivalent to the dissemination of the Bible. As the Bible was disseminated to more and more people, Protestantism reached these people. As the beginnings of the educational systems in England, the United States, and Germany are discussed, this relation will be shown.

In England, movements to counteract illiteracy were organized at the beginning of the eighteenth century. The Society for the Promotion of Christian Knowledge (SPCK), founded in 1698, intended that the child of the common man be taught to read the Bible and Christian writings such as the prayer book. Many in the upper classes opposed SPCK because the lower classes, by learning to read, would learn to think and therefore would neglect their work. Nevertheless, by the middle of the eighteenth century about fifty thousand children were in attendance at SPCK schools.

Again, in England in 1785 the Sunday School Society was formed for the purpose of teaching people to read the Bible. Within a decade, the Society distributed approximately twenty-five thousand Testaments and five thousand Bibles as sixty-five thousand pupils were trained in one thousand schools.[3] There were in England other educational efforts, usually carried out by Christian organizations with the same Bible-dissemination motive. In 1811 much of the burden was taken over by the Church of England. The nationalization and the universalization of education in England was actually a slow process,[4] but there is no doubt that the basic reason there was a movement at all arose from the desire to educate large numbers of persons who would consequently be able to read the Bible and that which was written about the Bible.

There was a similar pattern in the American colonies and later in the United States. In seventeenth-century Puritan New England schools were begun in order that children would be able to read the Bible. In founding these schools the Puritans were consciously anti-Roman Catholic, maintaining that the Roman Catholic Church was attempting to keep the Bible from the common man. The American colonies were later affected by an offshoot of SPCK, the Society for the Propagation of the Gospel in Foreign Parts, founded in 1701. This society also started schools so that large numbers of children would learn to read the Bible and material such as the catechism. Schools were begun by this society in all the colonies except Virginia, where it was thought they were not needed.

The Sunday School movement spread to the United States in 1786. In the schools begun there was great emphasis on religious instruction, with little secular instruction. These schools "helped point the way to universal education."[5] Other groups also founded schools. For example, in Pennsylvania there was A Society for Propagating the Knowledge of God among the Germans. One of the stated purposes of this society was the teaching of "the true principles of the holy Protestant religion."

The motivation for education during the early years of the United States was much the same as it was during the colonial period. Schools for small children became more common. Typical was the approach used in a Boston school founded in 1818, where children attended from the ages of four to seven. As they began this four-year school they learned the alphabet. They were to be able to read the New Testament by the time they completed their course. After the first years of the new Republic, much was still needed before there could be universal education. Yet, many were being educated, and the movement towards universal education was given much of its impetus by the need of the common man to gain ability to read the Bible.

In Germany educational growth began at the top. Thus, the approximate beginning dates of the various levels are the fourteenth century for the universities; the sixteenth for the secondary schools; and the sixteenth and seventeenth for the elementary schools, While there was at an early date lower-level education for those who eventually attended the secondary schools and universities, the elementary schools of interest in this discussion were those which were to be used by the common man. Up to the last part of the eighteenth century, the schools at all levels were ecclesiastically controlled, even though political leaders had often been active in founding schools.

Luther had insisted that the common man be educated. The Bible was no longer to be chained, and the common man was to be able to read it in his own tongue. Luther translated the Bible into German, and schools were needed to teach the common man to read it. Consequently, the elementary schools *(Volkschulen)* were an outgrowth of the Reformation. The motivation for the development of these schools was similar to the motivation for school development in England and the American colonies. This motivation is seen in the "General School Regulations for the Country," promulgated in 1763 by Frederick the Great. The Regulations stated that children were to attend school in order that they

> . . . know not only what is necessary of Christianity, fluent reading, and writing, but can give answer in everything which they learn from the school books prescribed and approved by our consistory.[6]

Thus, Luther's German Bible continued to have influence, and a principal factor in the eventual adoption of widespread education was the need of the common man to know "what is necessary of Christianity."

Small countries near Germany, England, and the United States developed in a similar manner. In Switzerland the educational change caused by the Reformation came through the work of Calvin. J. E. Wise, a Jesuit, said of both Luther and Calvin, "[man] had to learn in order to read Scripture."[7] Calvin emphasized higher education. He founded the Academy of Geneva, which later became the University of Geneva. There were fifteen hundred students in this school when Calvin died in 1564. Many of the Academy's students carried Calvinism to other countries, and as they did so, they founded schools at all levels. In Holland, for example, they organized grammar schools, vernacular schools, and universities, eventually founding the University of Leyden in 1575.

The Academy's influence extended to Calvin's native country, France, where by 1600 the Huguenots had founded thirty-two colleges and eight universities. The trends started by Calvin's followers in France were later reversed and the Huguenots were forced out of France. Consequently, France is not to be classed with Germany, England, and the United States in the matter of Bible dissemination. The monarchy was opposed to general education, and on the eve of the Revolution there was only a small number of elementary schools. The Revolution abolished all schools, regardless of the level, and alternately organized and abolished new systems. Napoleon organized a central school system in 1808, but even then there were only a few elementary schools. In 1833 Louis Philippe called for a large number of elementary schools, but it was not until the last part of the nineteenth century that these schools came into being. Consequently, France was well behind its neighbor, Germany, in the development of general education.

It was pointed out earlier that French scientific contributions lagged somewhat behind the contributions of the more productive countries. Without doubt, France contributed more to science than other countries which had not had the benefit of a widespread dissemination of the Bible. Yet, it seems that much of the French work came later, and that some of France's contributions would not have been made were it not for the influence of other countries. France was held back in the early years of the explosion of scientific knowledge. Consider the case of the chemist, Lavoisier, who was one of the greatest scientists of his day. Because of the anti-intellectual mood of the Revolution, an attitude which kept the masses from finally becoming educated after their long enslavement, Lavoisier was guillotined. This spirit hurt France's scientific effort during the critical years of the explosion of scientific knowledge, the years Germany, England, and the United States produced so much. These conclusions concerning France are surprising to those whose attention has been focussed on the arts, where the picture is quite different. No claims concerning a correlation between development of the arts and Bible dissemination are made here.

What of the rest of the world, outside those regions which have been labelled productive? Today there is intense scientific activity in many countries, such as Russia, China, Japan, and Israel, which are outside the initially productive areas. There is also mass education in these countries. The motivation for mass education came late, and it was a nationalistic, not a religious motivation. In each case it can be shown that mass education and scientific activity came *because* these activities had begun earlier in other places.

For example, in the decades prior to World War II Japan was known as the nation which progressed by copying what had been developed earlier in other countries. Similarly, revolutionary Soviet Russia inaugurated mass education with a heavy emphasis on science and technology in order that she be able to combat — both militarily and ideologically — countries which had already begun mass educational and scientific activity.

Up to now only the correlation between widespread Bible dissemination, mass education, and scientific activity has been discussed. The correlation could be fortuitous. However, there seems to be a causal connection, and this connection is discussed next.

Widespread Bible Dissemination Causing Scientific Activity

For two reasons science seems to have flourished wherever there has been dissemination of the Bible to the common man. The one reason is associated with the Bible being a book, and the other with the nature of the Biblical message.

A. *Why Does Being a Book Make a Difference?* The book must be read, and so the Bible-believer acquires the skill of reading. Reading brings with it education in general. The more literates there are in a population, the more there are who can be brought to the level of education which enables them to make scientific contributions. There is no reason to expect that the population of one country is more intelligent than the population of another country. It is rather a question concerning which population has the greater opportunity to develop. Thus, other things being equal, one may expect of Countries A and B of equal population, that if 1% of the people in Country A become literate, while 25% become literate in Country B, that Country B will produce many more top-ranking scientists.

There are at least three specific reasons for the connection between literacy and scientific activity.

(1) Only by reading can an individual be at least partially freed from prejudice. Without the ability to read, much of an individual's knowledge of the world is that which he can acquire first-hand. Consequently he is apt to entertain extremely distorted ideas about that which he cannot learn for himself. Illiterate persons seem to think and speak less reasonably than those who are literate.

Removing even a measure of prejudice tends to help scientific activity because the mind becomes accustomed to encountering the world as it is. One cannot build a world of fancy. When reading is nonexistent or at a minimum, old wives tales abound. Literacy connects people and enables them to accumulate and pool their knowledge. Isolated groups and individuals develop contradictory ideas about the real world, and when they develop such ideas, it is probable that no one idea is correct.

Naturally, the man who learns to read in order to read the Bible learns first of all about God, his relation to man, and whatever else God chooses to teach him directly. Such a person also reads outside the Bible, and some of what he learns is true information about this world. He learns that magic is trickery, that ghosts are fictitious, and that there is much more to the world than he had thought. As his horizons broaden, his mind is disciplined to stay with the facts as they are.

(2) Those who learn to read become accustomed to separating truth from falsehood. Often that which is written is either true or false. The nebulous, half-baked idea fares worse on paper than in the mind.

There have been many sharp theological debates and other kinds of debates in countries dominated by non-Christian religions, such as ancient China, Egypt, and Greece, and Islamic countries of later times. These debates were different from those which occurred where the Bible was disseminated widely. Where the Bible was widely read, large numbers of people became engaged in mind-sharpening discussion and debate. In non-Christian cultures few, often only the temple priests, were able to discuss and debate at length. The Christian and the non-Christian countries were not comparable.

There is an objectivity about the data the Bible provides. In the same way, there is an objectivity about the phenomena of nature. In neither case does truth depend upon the mind of the observer. (However, it is not proper to claim objectivity for our *descriptions* of natural phenomena; this is discussed in Chapter 4.) A similar statement about the arts cannot be made. Even though there are certain unchanging principles in the arts, objectivity does not play as prominent a role. Thus, it is possible that most of us living today would find Chinese literature of a thousand years ago to be rather boring. Yet, gunpowder and the compass, produced by the Chinese of that period, mean much the same to us as they did to the Chinese of that day.

As the minds of many are trained to work with objectivity in the Bible, there is introduced into the culture the ability to work as well with objectivity in other parts of creation.

(3) The mind trained to read thereby acquires a certain amount of self-discipline. He who learns to read has begun to learn to channel thoughts, and to stay with a self-appointed task. Some who develop this self-discipline are capable of scholarly work.

There have, of course, always been scholars, but it is a matter of simple arithmetic to show that there exists the possibility of a large amount of scholarly work wherever large numbers of people have learned the beginnings of mental self-discipline. Scientific activity is not *merely* the exercise of mental self-discipline. However, the imagination and the intuitive leap characteristic of the scientist are meaningless without mental self-discipline. The successful scatterbrained scientist is a myth. A necessary part of scientific activity is using logic in a most sophisticated manner.

The importance of logic in science can be illustrated by the structure of a mathematics or a natural science textbook. Even if one is unfamiliar with the subject matter of the book, he may be able to understand its first pages. If he then skips to the end of the book, what he sees seems to be an incomprehensible maze. Yet, if the book is well-written, the intervening pages develop the subject logically, step by step, so that after thousands of steps the last pages can be seen by the careful reader to follow from what precedes them.

Another example of the use of logic derived from the best kind of mental self-discipline is the laboratory synthesis of "natural products," which are very complex chemical compounds found in plants and animals. The chemist considers the successful completion of a single new chemical reaction, yielding a new compound, a distinct achievement. Some natural products, such as quinine, can be made only after dozens of *sequential* reactions are carried out successfully. Groups of organic chemists have labored for many years before achieving the goal of carrying out all the required reactions for only one natural product in the correct order. In recent years many natural products have been synthesized in the laboratory, and in each case the use of sophisticated logic was required.

The same devotion to mental self-discipline, with the consequent use of difficult logic, is necessary to determine the structure of a complicated crystal or molecule. A Nobel prize was awarded for the determination of the complicated structure of the DNA molecule, a key component in the living cell. The immense difficulty in untangling the structures of these complicated crystals and molecules can hardly be appreciated by the non-scientist. Without self-discipline and the use of logic, these structures would remain forever unknown.

Modern cultures which have produced many minds that have learned self-discipline by reading, have produced minds that could build scientific achievement upon scientific achievement, so that by now thousands of scientific problems, once thought hopelessly complicated, have been solved.

B. *How Has the Nature of the Biblical Message Affected Scientific Activity?* As the Biblical message was taught and accepted, the intellectual climate produced was favorable for scientific work.

Christians tend to be on the defensive concerning the intellectual climate brought about by acceptance of the Christian faith. Unbelievers have repeatedly stated that the *church* has stifled science. Over and over they have cited the church's opposition to Copernicus' overthrow of Ptolemy's cosmology and Galileo's proof that Aristotle erred. Unbelievers extend this attack to the Christian church in general. What is maintained here is that *the widespread dissemination of the Bible*, not a church, has provided the climate so fruitful for the growth of science. Where the Bible was kept from the people, their intellectual activity, as reflected in the development of their educational system and the level of their scientific activity, was poor. Except for twentieth century developments, Roman Catholic countries, especially prior to the Reformation, have not nurtured sustained scientific effort.

Even though there was endless strife and much sin in the productive countries during the critical period of the explosion of scientific knowledge, men were not stupefied, shackled slaves, without lofty ideals. Three uniquely Christian factors were important.

(1) Men served God more and men less. Even though monarchies

did not end, the spread of the Christian gospel gave many men a degree of independence. The humblest peasant, though he remained a peasant, was lifted up because he knew he was personally and directly responsible to God, and because he knew that God's blessings came through *no* human intermediary. The common man who accepted the gospel received a new dignity. He knew he had a mind. It was easy for all common men to catch something of this spirit, although many, perhaps most, did not accept the gospel. Oppressors may have remained unconverted and oppressors still, but life changed permanently when many common men came to realize that man was created in the image of God. The mind of man could no longer remain undeveloped. Science became inevitable.

(2) Another factor which helped scientific activity was the realization of many that the physical universe is important because God created it. The artificial division between nature and grace was gone for those who accepted the Reformation principles. Nor could the Reformation man possibly believe that evil is to be identified with matter and good with spirit, even though many people had earlier uncritically accepted these ideas. God's instruction to subdue the earth (Gen. 1:28) was a message not only to the few who had the time and money to obtain an education and investigate; this instruction was given to all believers. Those who believed the Bible learned that God created a universe which man was to develop. Developing and understanding the physical world came to occupy the lives of many, both believers and unbelievers. Unbelievers were affected by the new ideas and a large number of people eventually became engaged in scientific work.

(3) Science developed in many different directions. One of the most important areas which developed was the study of the human body and whatever else was needed to aid in the preservation of human life. With the Reformation-born understanding of human dignity there came an emphasis on the value of human life. Even today human life is cheap in countries in which the Bible was never widespread.

Much of the explosion of scientific knowledge can be traced to the growth of the sciences related to health. It was inevitable that human comfort, not merely health, would become the goal of scientific effort. "Better living through chemistry," "better living through physics," etc., express an important motivation for much scientific work. This motivation is now associated with the very opposite of the Christian faith, materialism, the philosophy of the natural man. What is often not seen is how modern scientific activity began. The idea that health and comfort are to be cherished is a Biblical idea.

The thesis of this chapter along with the reasons used to support the thesis ought to indicate to all, including the modern scientist, that science as we know it would not exist were it not for the impact of the Bible on certain populations at certain times. Naturally, the Bible claims the allegiance of men irrespective of recent history. Yet, God in his providence caused the Bible to be a major factor in recent history. Therefore, it seems proper to use this historical fact to call the Bible to the attention of those who have a high regard for modern science.

REFERENCES

1. Asimov, I., *Asimov's Biographical Encyclopedia of Science and Technology,* Doubleday, Garden City, N. Y., 1964; pp. 328-330.
2. Mulhern, J., *A History of Education – A Social Interpretation,* Ronald Press, New York, N. Y., 1959; p. 525.
3. Graves, F. P., *A Student's History of Education,* Macmillan, New York, N. Y., 1918; p. 237.
4. Graves, F. P., *A History of Education in Modern Times,* Macmillan, New York, N. Y., 1913; p. 301.
5. *Idem,* p. 238.
6. Quoted from the Regulations in Ref. 4, p. 281.
7. Wise, J. E., *The History of Education,* Sheed and Ward, New York, N. Y., 1964; p. 207.

3 *The Bible and Scientific Investigation*

For centuries Christians have claimed perfection for the Bible. Prior to the present era, the term "perfection" itself may have been imperfectly defined by many scholars who used it. In the present scientific age, because of a wide use of the analytical approach, many old concepts are being scrutinized and examined more carefully than ever before. So, too, the concept of Biblical perfection is being analyzed from every conceivable angle.

Much of the present interest in science has arisen, as was shown in the last chapter, because some men believed the Bible to be perfect. Today scientific analysis comes back to the Bible and asks if the Bible does indeed possess perfection. Sometimes a man returns to visit the fondly-remembered teacher of his student days, only to discover that he now far surpasses his former teacher. Is this the situation with modern science and the Bible? Does modern scientific analysis show that the Bible is not perfect?

It is no coincidence that scientific analysis should now be applied to the very book which once gave men new motivation to analyze both that book and nature. Scientific analysis, or scientific work, implies a search for truth. If the Bible contains *absolute truth,* which the scientist never finds outside of the Bible, there could be amazing implications for natural science scholars and other scholars. Historians spend countless hours searching ancient libraries to obtain correct historical information. Some scholars are occupied their entire professional lives in thinking, reading, and experimenting, for the purpose of developing correct information about animals, plants, minerals, stars, or anything else in the universe. Scholars willingly assume that they might spend their professional lives producing no more than a few pieces of correct information. Is it possible that the Bible, with its hundreds of pages, contains statements which historians, scientists, and other investigators may *absolutely* rely upon?

Questions Concerning Biblical Perfection

Throughout this book an effort is made, as explained in Chapter 1, to show that the natural man and the Christian interpret their environment differently. (Whenever references are made to the Christian position or to the natural man's position, it is to be remembered that these are idealized positions, and not the positions of specific persons, who tend to be inconsistent in adopting some of each of these two positions.) Consequently, the natural man and the Christian view the Bible differently. For the natural man, who maintains that only "nature" is real, the Bible is just another part of nature. For him, it is a book written by men, a book fundamentally no different than other books written by men. The natural man observes that men are prone to error, and he concludes that therefore the books of men, including the Bible, can contain error. The natural man does not claim perfection for the Bible.

It was pointed out in Chapter 1 that the Christian position is a meaningful alternative to the position of the natural man because the Christian and the natural man differ sharply on whether or not there is a god outside this universe, a sovereign god who created the universe. In that discussion it was stated that the Christian position also implies an acceptance of the idea that the Bible is God's revelation to man. The role of the Bible in the Christian position was not discussed further in Chapter 1; this matter is taken up in this chapter. In contrast to what the natural man believes about the Bible, the Christian who consistently holds to the Christian position maintains that the Bible is perfect. Concerning this belief, the scientist has several questions. These are now considered.

A. *What Is It That Is Perfect?* The Christian position is that the original manuscripts of the books of the Bible were perfect, or inerrant. Insofar as our versions faithfully reproduce these manuscripts, they too are inerrant.

At first it seems meaningless to claim inerrancy for the original manuscripts, documents which no longer exist. But it can be shown that the original language texts which present-day scholars depend upon correspond almost perfectly to the original text. There are two ways the reliability of the generally accepted texts can be demonstrated.

One way is to consider what is said by the scholars who study ancient texts of the Bible. They have found that we would have the same Bible, almost word for word, if only some, but not all, of the ancient texts were used. Thus, there are independent lines of successive texts, and almost the identical Bible can be derived from the different lines. There is therefore almost certain knowledge as to which words of the generally accepted texts are in question. It is also known what the various possibilities are for these uncertain passages. There are only a few hundred words in question out of about three-quarters of a million words in the English text. Theological differences of opinion which have arisen have not arisen because of these few hundred words. Those who claim the Bible errs do not rest their argument on these few hundred words. No part of the Christian faith would change were these words understood differently.

The other way the generally accepted texts can be known to be nearly perfect is by a consideration of the promises of God. He promises in these texts that we today can use the Bible for instruction.

> All scripture is given by inspiration of God, and is profitable for doctrine, for reproof, for correction, for instruction in righteousness. (II Timothy 3:16)

Christ instructed the apostles to give his commandments to all nations.

> Teaching them to observe all things whatsoever I have commanded you: and, lo, I am with you alway, even unto the end of the world. (Matt. 28:20)

He also said,

> Heaven and earth shall pass away, but my words shall not pass away. (Matt. 24:35)

These and other passages either imply or state directly that God's Word will always be with us and that we can rely upon it.

If the generally accepted texts did contain significant deviations from the original, not all of the Bible would be dependable. We could not, referring to the passages just cited, receive Christ's commandments. Either the generally accepted texts are virtually perfect, or these promises concerning the future of the Bible were not made. If such promises were not made, there is no Christian faith left. But the Christian knows that the faith as taught in the Bible is true. Thus, the choice is between the Christian faith and an unreliable text: there is no middle road. Only those who believe the Christian faith to be fundamentally wrong can consistently deny the reliability of the generally accepted texts.

B. *What Is Meant By Inerrancy?* Various kinds of error might exist in a book purporting to be the holy book of a religion. To clarify the use of the term "inerrant," some of the possible errors are listed and discussed.

(1) The holy book might teach incorrectly how man can meet with God's favor.

(2) The holy book might give incorrect information about life after death.

(3) The holy book might give incorrect ideas about the nature of God.

(4) The holy book could record the acts of God incorrectly.

(5) The holy book could record human history incorrectly.

(6) One part of the holy book might interpret another part incorrectly.

(7) The holy book might incorrectly describe the physical environment incidental to events recorded.

(8) The holy book might provide incorrect answers to scientific-philosophical questions, such as questions concerning whether or not the universe is finite with respect to age and size.

The meaning of Biblical inerrancy is that *the Bible cannot make either any of these errors, or any other error which might be supposed.* There are difficulties in understanding the Bible, difficulties which arise from the reader's limitations and not from an inherently clear Bible. At times the words have meaning only as they are understood in the context of the culture in which they were spoken or written. Regardless of how much we must cast off our culture to understand the Bible, and regardless of the difficulty of reconciling supposedly irreconcilable passages, there is one conclusion which cannot be made, namely, the conclusion that the Bible contains error.

C. *What Is the Proof of Biblical Inerrancy?* Attempting to prove Biblical inerrancy from the Bible itself seems at first to be circular reasoning. Yet, proving the inerrancy of the Bible from the Bible is just the procedure which will be used. To justify this procedure, consider the answer to the question, "If the Bible is inerrant, how could it be known to be inerrant?"

There are apparently two ways whereby Biblical inerrancy could be known. In one way, God could give us private revelation, telling each person that the Bible is inerrant. Christians do not claim to have received these private revelations. They do not maintain that they know of inerrancy because of dreams, visions, and other private revelations.

The other way Biblical inerrancy could be known is by the testimony of others. For example, God might have revealed the fact of inerrancy to ancient writers. Ordinarily, one would suppose that ancient writers were fallible and not to be depended upon any more than anyone else. Yet, if an ancient writer did indeed learn from God that the Bible is inerrant, *his writing itself becomes a part of the Bible.* Thus, there cannot be reliable, *extra-Biblical* testimony concerning the inerrancy of the Bible, and if there is reliable testimony concerning the inerrancy of the Bible, it will be found in the Bible itself. Thus, the Christian position is that God did indeed tell various ancient writers that the Bible is inerrant, and that God, through his Holy Spirit, instructs the individual Christian to accept their witness. Therefore, one *must* turn to the Bible to learn of inerrancy. There is no other place one could possibly learn of Biblical inerrancy.

Three of the Biblical passages in which inerrancy is taught and defined are discussed here.

(1) Jesus said,

> For verily I say unto you, Till heaven and earth pass, one jot or one tittle shall in no wise pass from the law, till all be fulfilled. (Matt. 5:18)

An incorrect law would not be fulfilled. Thus Jesus claimed that the law is correct. He did not limit its correctness to the general idea of phrases and sentences. It is correct down to the (approximate) English equivalent of the dotting of the "i" or the crossing of the "t."

(2) Jesus also said, quoting Exodus 3:6,

> Have ye not read . . . I am the God of Abraham, and the God of Isaac, and the God of Jacob? God is not the God of the dead, but of the living.
> (Matt. 22:31-32)

Jesus was refuting the Sadducees, who did not believe in resurrection.

They would have said that Abraham, Isaac, and Jacob no longer existed when God spoke these words to Moses, who lived hundreds of years after they lived. By using the Bible in this way, Jesus teaches the correctness of the Bible. His answer would have meant nothing if it was possible that the Bible contained error. He also teaches that the correctness applies to the smallest detail, since his whole argument rests on the tense of the verb in the phrase, "I am." Because God said, "I am," and not "I was," God was saying that Abraham, Isaac, and Jacob lived, though they had died.

Interestingly, in these passages Jesus referred to the version of the Old Testament available in his day, not to the original manuscripts.

(3) In the passage most frequently used in discussions concerning Biblical inerrancy, Paul says,

> All scripture is given by inspiration of God, and is profitable for doctrine, for reproof, for correction, for instruction in righteousness. (II Tim. 3:16)

The Greek text is more forceful: scripture is "God-breathed," and therefore the Bible is so perfect, and so free from error, that *all* of it (referring here, of course, to the Old Testament) can be used for our profit.

Many problems concerning inerrancy have been raised by scholars. There are, for example, several pairs of apparently contradictory Biblical passages. Scholars can in some instances understand how these contradictions *may* eventually be reconciled. It is not as important as is sometimes thought that there be such a reconciliation. What is important is that *man has no means of finding a contradiction in the Bible.*

How can such a sweeping claim be made? Suppose there were a means of determining that there are contradictions in the Bible. For example, one might hold that "common sense" could be used as a yardstick to ascertain whether or not passages are contradictory. Whatever yardstick one would use, however, the use of the yardstick could not be limited to certain passages. The Bible does not indicate that certain passages may be judged, while others are to be left alone. Therefore, any element of Christian theology could be evaluated with the use of such a yardstick, and Christian theology would then ultimately depend upon this man-made yardstick. The Christian position is that Christian theology cannot depend ultimately upon man. Therefore, a yardstick whereby contradictions can be found does not exist.

The inability of man to find a yardstick is taught by II Timothy 3:16, quoted above. This passage states that "all scripture" is profitable for us and it does not permit an inerrant yardstick, another authority. There cannot be two ultimate authorities, two courts of last resort.

D. *Can Scientific Truth Be Found in the Bible?* Even if the Bible is inerrant, the question concerning whether or not inerrancy has meaning for the scientific investigator must be answered.

Frequently it is said, "The Bible is not a science textbook. Therefore, one cannot expect to find science in the Bible." Such a statement is not a claim that the Bible is in error. It is a claim that the contents of the Bible are not relevant for scientific investigation.

A statement which declares that the Bible does not contain science is a statement *about* the Bible, and such a statement is not derived from the Bible. A very important rule which is used by many (perhaps most) Bible scholars is that the only true statements about the Bible are those which can be derived from the Bible. The truth of this rule is accepted in this discussion, but it is not enough to accept it on the authority of Bible scholars.

To prove this rule, it will be temporarily assumed that the opposite of the rule is true, and the consequences of such an assumption will be examined.

Assume that one is able to formulate true statements about the nature of the Bible which cannot be deduced from the Bible. It is always in principle possible that such statements could at least in part determine the correct interpretation of the Bible. In fact, it is difficult to conceive of statements about the nature of the Bible which could not affect interpretation. Because these statements could affect interpretation of the Bible, they would in effect be put on a par with Biblical statements. Since such addition to the Bible by man is not permitted, these statements about the nature of the Bible cannot be known to be true.

The error in asserting the truthfulness of statements about the nature of the Bible is a subtle error. There is a tendency by those who accept inerrancy to fall into this error. Once the inerrancy of Biblical statements is accepted, it apparently is easy to postulate the truthfulness of a characterization of those statements, even if that characterization cannot be derived from the Bible.

There are many statements about the nature of the Bible which are correct because they can be deduced from the Bible. Thus, the Bible is said to be inerrant because this concept is taught by the Bible. The Bible teaches that it is the Word of God. It is the Bible that teaches that the whole Bible is to be used for instruction in righteousness. Likewise, it is deduced from the Bible that God's salvation in Christ is at the heart of all of the Bible.

The statement, "The Bible is not a science textbook," is deficient because it cannot be deduced from the Bible. Similarly, the statement, "The Bible is a textbook of . . ." is also deficient. The Bible cannot be limited by any statements such as these. The Bible teaches what it teaches. For example, it is a mistake to expect that the book which explains God's plan of salvation would not at the same time give us a correct biological fact. One Biblical statement or set of statements which seems to outline God's plan of salvation cannot be used to predict correctly the contents of another Biblical statement or set of statements.

An example of our inability to determine from part of the Bible what it teaches in another part is provided by comparing certain passages which include the story of Rahab. The ninth commandment says, "Thou shalt not bear false witness against thy neighbor" (Ex. 20:16). The reader of the account which describes how Rahab hid the spies sent to Jericho (Josh. 2:1-7) might conclude that Rahab broke the ninth commandment. Yet, the Bible teaches that *this act* proved Rahab to be among the saved:

> Likewise also was not Rahab the harlot justified by works, when she had received the messengers, and had sent them out another way? (Jas. 2:25)

A similar thought concerning Rahab is expressed in Hebrews 11:31. The James and Hebrews statements do not contradict those in Exodus and Joshua. Rather, the Biblical teaching on bearing false witness cannot be completely understood until all of the Bible is permitted to speak.

Sometimes it is claimed that a certain scientific idea cannot be found in the Bible because it is not in the "character" of the Bible to discuss scientific concepts. This claim is supposedly supported by the Bible itself:

> All scripture . . . is profitable for doctrine, for reproof, for correction, for instruction in righteousness. (II Tim. 3:16)

Are not scientific matters, it is asked, quite different from doctrine, reproof, correction, and instruction in righteousness, all of which are related to the Christian faith?

Consider the problem of the Bible-believing historian. He might initially suppose that his faith and his scholarship are separate. Yet, there is a myriad of historical facts in the Bible. They are *all* profitable for doctrine, for reproof, etc. They are *all* correct facts. The Bible-believing historian is thus not able to separate his discipline from his faith. In the same way, any fact in the Bible could have a bearing on one or more disciplines, including the scientific disciplines. Men are not permitted to limit God's meaning when he speaks to them.

Unfortunately, the idea that the Bible can contain scientific fact has occasionally been accepted in a rather superficial manner. Some persons can supposedly find in the Bible prophecies of many scientific achievements, such as the development of the automobile, the airplane, and atomic energy. Such interpretations of the Biblical text are easily refuted. But the approach taken in this book is that such ideas should not be refuted with the claim that the Bible does not deal with such things. A refutation of any such interpretation may be made on the basis of a careful study of the Biblical text. In this way, making an improper statement about the Bible is avoided.

E. *Can the Pre-Scientific Biblical Statements Be Related to Science?* In spite of the arguments of the previous section, it is widely held that Biblical statements cannot possibly have relevance in scientific matters because the Bible was written by men who knew no science to men who knew no science. The Bible is pre-scientific. Without question, this is the most frequently given reason Christians have for keeping the Bible out of scientific discussions.

If this reason is valid, then certain Biblical prophecies do not mean what Christians think they mean. David, Isaiah, John, and many other prophets surely made some prophecies which neither they nor their first readers fully comprehended. Yet God guided them to write correctly. If the Biblical authors could not have written science because their culture was pre-scientific, then neither could they have written prophecies which are properly understood only when they are fulfilled.

Consider, however, the example often used to prove that Biblical statements cannot have scientific validity. To prove the point, it has been suggested that the Bible teaches that our universe is a three-story universe. Heaven is in the skies and hell is below the surface of the earth. It is presumed that this concept was commonly held in Bible times, and that it crept into the Bible. Some Christians claim that these are but "tangential" statements, and that it is not necessary to suppose that the error of these statements in any way affects our faith.

Paul in Philippians is said to have alluded to the idea of a three-story universe as he alluded to the exaltation of Jesus:

> That at the name of Jesus every knee should bow, of things in heaven, and things in earth, and things under the earth. (Phil. 2:10)

Three aspects of the three-story problem are considered.

(1) Suppose that the three-story concept was held by Paul, that Philippians 2:10 reflects this belief, and that his early readers also had this belief. The three-story concept was, however, at a later time proved false. If the supposition is correct, those who proved it false could not have used the Bible in the proof. Whoever was the first to claim that the three-story concept is false would have contradicted that which men believed the Bible taught. *Up until that time, there would have been no way to proclaim the supposed three-story implication of Philippians 2:10 to be a tangential concept.* Only after proving the three-story concept false could a three-story implication of Philippians 2:10 be labelled tangential. Thus, if there is a three-story implication of this verse, it is not possible to decide which other Biblical statements are tangential until science, secular history, etc., are applied. It appears that a consequence of this supposition is that only extra-Biblical information can indicate which statements are tangential and not relevant to the Christian faith.

(2) It has, however, been suggested that the three-story implication

can be known to be a tangential matter because such an implication of the text has no bearing on our faith.

Consider the position of the Bible-believers who lived before the scientific age. They are supposed to have accepted the three-story implication of this passage. These believers would have been prevented *in principle* from predicting our modern space programs. It would have been wrong for some far-sighted person to visualize the combination of Newton's laws of motion (first given in the seventeenth century) and combustion to produce man-made satellites and other space probes. If the Bible had seemed to tell men that God lived in the sky, it would be wrong for them even to contemplate the day when men would travel to his abode via a space ship.

Perhaps it is even easier to imagine an early reader of Philippians 2:10 contemplating the possibility of digging a giant hole in the earth. Even contemplating such a project would be wrong if hell were there. Yet, men do now consider digging a hole large enough to penetrate the earth's crust ("Project Mohole") under the Pacific Ocean. Even though an early reader would have been able to visualize this quite-feasible project, it would have been wrong to do so if the Bible seemed to say that hell is below the surface of the earth. Because the acceptance or non-acceptance of the three-story implication of Philippians 2:10 was thus an *ethical* matter, this supposed implication is not tangential. A three-story implication of this verse does have relevance for the Christian faith, and therefore one cannot correctly maintain that the statement can be known to be tangential for the reason that it is not relevant to the faith.

Note that the objection to an early believer's making these predictions about modern science does not arise because he is assumed to have interpreted Philippians 2:10 incorrectly. The assumption for the sake of this argument was the assumption that these early believers *correctly* interpreted the passage, and that any error was the error of the writer.

What Philippians 2:10 *does* teach is that no part of creation is beyond Christ's authority. No man, whether he is a scientist or any other kind of explorer, will ever be able to maintain that he has found a place where Christ does not rule.

What has been shown for the three-story concept in Philippians 2:10 could, in principle, be shown for other supposed tangential statements. No one can say that any Biblical statement is a tangential statement, even though it may at present seem to be such a statement.

(3) Today we know from science that the universe is not a three-story universe. It may well be that men once thought that the Bible teaches a three-story concept, but if so, it may now be concluded that their understanding of the Bible was imperfect. Scientific knowledge

can help us understand the Bible. Science can tell us that early believers read Philippians 2:10 incorrectly.

It is possible to conceive of a situation in which men are wise and very close to the Lord, so that they could interpret all of the Bible correctly without the aid of scientific investigation. In real life, however, men very easily come to wrong conclusions, and so it may be that historical and scientific research will help remove our inabilities in understanding the text. The best research tools ought to be brought to bear on exegetical problems. Bible scholars profitably use many methods to understand words and phrases correctly.

As Bible scholars attempt to obtain precise meanings of Bible words and phrases, there is always the danger that they might detract from the perfection of the text. Van Elderen says,

> The doctrine of organic inspiration maintains that God used these authors in their settings and environment with their knowledge, culture, cosmology, view of reality. We must recover these elements in order to understand their writings and often archaeological and historical research are the means of recovery.[1]

Recovering the cultural setting of the author is important in order that his words and phrases be properly understood. Yet, of what use are his *beliefs* – for example, his cosmology, mentioned in the quoted passage – if these beliefs are not approved by the Bible? Suppose it is proved that Paul's contemporaries accepted the idea of a three-story universe. Suppose one argues from the general to the specific and concludes that Paul himself held to this concept. It does not follow that Paul would have recorded in the Bible an erroneous idea he had. Even though it is the best kind of approach to the Bible to attempt to understand it by understanding more of the cultural setting of the writers, it does not follow that error can thereby be found in the text. There is no logical way to show that error in the text is a necessary consequence of error in the mind of the author.

When Jesus spoke of a camel going through the eye of a needle (Matt. 19:24), he very likely meant by a "needle," as scholars assert, something quite different from what we in the Western world would mean if we used these words. Here is an example of how it is possible to learn more about the Bible by learning of its cultural setting. In such examples, learning of the cultural setting does not make it legitimate to charge the text with error. After the cultural-setting argument is thoroughly examined, there should be no difficulty in accepting the idea that God kept error out of the Bible, even though superstition and scientific error abounded in Bible times.

F. *How Can It Be Shown That the Bible Can Aid the Investigator?* In what preceded, it was maintained that the Bible is without error and that any scientific statements found in the Bible are relevant today. Although it was not shown that there are such relevant statements, it is possible to list several such statements. However, a different approach will now be used. Three examples of Biblical statements which can be, or could have been at one time, meaningful for the investigator are given; only one of the examples is related to science. (Some scientific statements which could be given now are instead taken up in later chapters, where they are the basis of more detailed discussions than can be given here.)

(1) Daniel 5 states that at the time Babylon fell, Belshazzar was king. In the early twentieth century some scholars said that this was an error, and that Nabonidus was in fact the king. They knew of no one named Belshazzar. However, by means of archaeological research it was discovered that Nabonidus had been the king over a large empire, of which Babylon was a part, and that his son, Belshazzar, was king in Babylon. It is thus explained why Belshazzar, second in command, rewarded Daniel (Daniel 5:29) by placing him third, not second, in the kingdom. Those scholars who believed the Bible knew a historical fact before their unbelieving colleagues knew it.

(2) Luke 2:2 refers to Cyrenius, also called Quirinius, who was governing at the time Mary and Joseph were to go to Bethlehem to be taxed, at the time Jesus was born. For a long time many scholars said that the reference to Quirinius was an error. They said that Quirinius was not governor until 6 A.D., while Jesus was born in 4 or 5 B.C. It seemed that Quirinius did not have authority over Palestine when Jesus was born. However, certain inscriptions have been discovered which have vindicated Luke's account. In the years after 7 B.C., at least up to the time Mary and Joseph went to Bethlehem, Quirinius served under Varus, the governor of Syria. Quirinius had authority in the Roman provinces over which Varus was governor; Quirinius was in charge of foreign affairs, and he would have been in charge of the taxation in the province of Palestine. Luke is very precise in this matter: the Greek text of Luke 2:2 does not say that Quirinius was **governor;** rather, Luke said that Quirinius was **governing,** exactly what would be said of the Syrian governor's subordinate who had authority in a province. Some scholars were therefore wrong about Quirinius' role in the events of 4-5 B.C. Here also Bible-believing scholars knew a historical fact first.

(3) There are in Leviticus 13 and 14 health rules which, if followed, prevent the spread of leprosy. These are sound rules, quite consistent with what is now known about contagious disease. It is not claimed that such safe practices were unknown among nations around Israel. What is important is that there have been many cultures in which this kind of hygienic care was not exercised. Hygiene is now widely practiced be-

cause of nineteenth and twentieth century scientific discoveries. Yet, certain of these practices could have been used in earlier centuries in more places than they were used, had the perfect and useful instructions in the Bible been accepted. This example indicates that correct scientific information can be found in the Bible.

These three Biblical statements have been confirmed through the use of modern research tools and methods. There are other factual statements in the Bible which are widely denied by scholars. If we see ourselves not at the end of scientific and historical research, but living while such research is still being carried out, we may realize that these "conflicts" are no more real than the Belshazzar "conflict" was real a century ago, before further research resolved the matter. On the other hand it is not necessary to expect that all conflicts will be resolved before the Lord returns.

The Biblical statements concerning Belshazzar, Quirinius, and the Levitical health rules are not statements which, when proven true by extra-Biblical investigation, cause much of a stir. The natural man is not very disturbed as he admits, for example, that he erred concerning Quirinius. For the natural man, the Bible is a part of the nature he investigates. According to him, the Bible, as one more historical document, will occasionally be correct when the other available sources are wrong. But many of the Biblical statements concerning scientific matters, some of which are discussed in later chapters, are certainly not trivial. If the natural man admits the Bible is correct where challenged on these "small" historical and health matters, he will have difficulty in challenging the Bible anywhere. On several of the scientific questions discussed later, questions about which the Bible makes definite statements, the natural man cannot afford to yield.

REFERENCE

1. Van Elderen, B., *Calvin Theological Journal, 1,* 165 (1966).

4 Natural Law and Miracle

There are two kinds of scientific information in the Bible. First, there is information which is usually considered non-controversial. For example, the hygienic care of lepers is described. Plants and animals which lived in Bible lands and in Bible times are identified, and such information can also be considered scientific information in certain contexts.

The Bible also provides information useful in answering certain profound scientific questions, such as, "Did matter have a beginning?", "Is the universe infinite?", and "What is the relation between miracles and natural science?"

To discuss miracles and natural science, the concept of "natural law" must be explained. It will be seen that there is a man-centered aspect of natural law, and that this subjectiveness extends even to the act of observation. After the quantitative aspect of natural law and the relation of natural law to time are explained, miracles will be seen to be part of a law of God which is broader than man-centered natural law. It will be shown that there is no sharp line of demarcation between miraculous and non-miraculous events, and that the only law which has perfect predictive properties is the law which describes both kinds of events.

Natural Law

Objects are observed to fall to the earth. If nearby objects are suspended on strings, the objects are observed to be attracted to each other. If a slow-moving space probe approaches the moon, the probe is observed to fall to the moon. These observations can be used to formulate the general statement, "Any two objects are attracted to each other." This generalization is one way of expressing the law of gravitation.

Such a generalization is the beginning of a description of our universe. As the scientist brings together many such generalizations, he formulates a "picture" of the universe. This picture is called a "model" of the universe. If this picture were a painting, each of these generalizations would be a brush stroke.

The model concept is also used with respect to the small segments of the universe the scientist usually studies. Note that the statements of which the model is comprised are always generalizations of specific events, such as an object's falling to the earth or to the moon. Even though the scientist cannot observe all possible cases, he generalizes and thus states the law of gravitation.

Each of these generalizations is a *natural law*. Some natural laws are related to observation in a more subtle way than the law of gravitation is related to observation. For example, in no process has it been observed that energy is created or destroyed; therefore, it has been concluded that energy (or its equivalent, mass) is never created or destroyed. The law of the conservation of energy is thus also a natural law, even though it rests not upon observation of the process which occurs, but rather upon our inability to observe the opposite of this law. In what follows, it is assumed that natural laws can rest upon either kind of observation. Four comments concerning natural law should be made.

A. *Man-Centered Natural Law.* Natural laws as defined above are formulated by man. In no case are these laws anything other than man's generalizations. Emphasizing the man-centeredness of natural laws does not detract from them, but it does help to place them in proper perspective. With this perspective the natural law concept can achieve its greatest usefulness.

Suppose, however, that the scientist formulates a *true* description of a segment of the universe he studies. Are the laws of which this description is comprised still *only* man-centered laws, that is, laws formulated *merely* by man?

Part of the answer to this question lies in the precise meaning of "true description." The scientist imposes a limitation on this phrase which the non-scientist might think unnecessary. By "true description" the scientist means that his law makes correct predictions about what can in principle be observed in as-yet-unobserved situations. If the scientist believes the law, "Any two objects are attracted to each other," to be true, he then expects that for any two objects, whether they are on our planet or billions of light years apart, there is observable mutual attraction.

The correct predictions which the law makes are *only* of that which is *observable.* The application of natural law to inherently unobservable situations is meaningless. (The discovery that certain phenomena are inherently unobservable, revolutionizing scientific thought, is discussed in Chapter 5.) A particular natural law is derived from observations and its correctness is tested by comparing its predictions with new observations. The law is useful to the extent it predicts what has not yet been observed, but it may not be used to predict that which cannot be observed. The scientist does not dare to take one step outside of the observation-law-observation circle.

There is another matter to consider in deciding if some laws are not man-centered. The laws which man formulates are limited by man's finiteness and imperfection. As the scientist applies a law, he cannot claim that the law is true in the absolute sense. He can, however, believe the law to be true in all observable situations. But the law can be no more reliable than the finite, imperfect man who formulates it. A law might be formulated incorrectly because an insufficient number of observations has been made. Observational error might creep in. The validity of a generalization might be ascribed to a broader class of phenomena than is justified by the limited number of observations which are made. The history of science affords many examples of each of these shortcomings.

It is sometimes tempting to ascribe absoluteness to a natural law because the law cannot be conceived to be untrue. Yet, inability to conceive of a different situation is insufficient reason to reject the possibility that such a situation exists. Consider the following analogy. If there were only one human language, men would probably ascribe some kind of absoluteness to the structure of that language. They would very likely be unable to conceive of the possibility of communication by means of another language. Children who have known only one language often express disbelief when they are told that far-away peoples speak differently. Yet, even if man could conceive of only one language, it would be foolish to place such a limitation on God: he could cause people to speak different languages. In a similar way, the description of the universe which man makes is not necessarily the only possible description. A natural law which man formulates is not an absolute law.

Unfortunately, the term "natural law" seems to imply absoluteness. The term suggests that "nature" functions according to laws which are independent of man and which man discovers. There *is*, however, an absolute law, a law which does not depend upon man in any way. This law is expressive of the will of the sovereign God. The correlations or generalizations, the natural laws which man formulates, do not govern creation. Creation is solely and completely controlled by the will of God revealed in his absolute law.

B. *Observing Facts and Formulating Laws.* The process of developing a natural law consists of two steps. The first step is the observation of events or situations; i.e., "facts" are observed. The fall of an object to the earth is a "fact." In the second step, a generalization is derived from the observed facts and a law is formulated. A common error is the assumption that observation is reliable, while generalization is unreliable. The "fact" is mistakenly thought to be beyond dispute, while the law is correctly assumed to be subject to re-formulation.

This misunderstanding concerning reliability arises because observation and generalization are thought to be qualitatively different activities. But a close examination of this question reveals that these two activities are not qualitatively different. An illustration will demonstrate this point.

Consider two metal balls, each electrically charged, placed near each other. The only difference between these particular balls is that one contains an excess of negative charges, while the other contains an excess of positive charges. An observer notices that these two balls are attracted to each other: if allowed, they move towards each other. The positive-negative attraction is said to be an observed *fact*. If one were to observe many positive-negative charge systems, he would in each case observe attraction. Further examination of these systems reveals that the fundamental negative charge is the electron and the fundamental positive charge is the proton. The generalization or natural law which emerges is, "Positive protons and negative electrons always attract each other."

When, however, one states that the positive proton and the negative electron attract each other, what is the difference between the observed fact and the natural law? It is impossible to conceive of a negative electron-positive proton pair in which there is *not* attraction. The law of electrical attraction is bound up in the nature of these two particles. To speak of these particles is to imply the existence of the electrical attraction law.

The example may seem to be improper because an appeal has been made to the natures of the negative electron and the positive proton. But it is the *nature* of the negative electron to be attracted to any positive particle; or, it is the *nature* of the positive proton to be attracted to any negative particle. Therefore, positive protons and negative electrons must attract each other. There is no other possibility. Any law-formulation process is basically the same as the one just given. Even though the scientist cannot often make an observation as simple as the observation made in the positive-negative attraction example, his complex systems always reduce to a few simple laws, each similar to the simple electrical attraction law in certain respects. In each case, whatever "obeys the law" – charged particles, masses, light, etc. – acts according to its own nature, even as it is the nature of the negative electron to be attracted to whatever is positively charged.

When the nature of the particle (or whatever else obeys the law) is known, the relevant law is also known. Natural laws and the natures of the particles, etc., which obey these laws are inextricably bound up together. In a fundamental way it is not possible to separate observation of fact and the formulation of natural laws. Expressing this, van der Laan said, ". . . One encounters the law side and the factual side of reality as correlatives."[1]

There can, however, be error in the formulation of a natural law. Since the observation of facts and the formulation of natural laws are not qualitatively different activities, it follows that fact observation can also be in error. Without doubt, those events or situations termed "facts" are more reliably known than the natural laws which are formulated. Yet, there is uncertainty in both facts and laws for the same reason.

There is another, entirely different approach which shows that facts and laws are uncertain for the same reason. Suppose in a very simple system of investigation that there is a box on a table in a well-illuminated room. An observer in the room concludes there is a box on the table because he can see it. Suppose that the observer turns around, so that he cannot see the box directly, even though it can be seen by means of a mirror. There can be another arrangement in the room, such that the observer cannot see the box except with the aid of *two* mirrors. This process can be continued, so that the observer cannot see the box without the aid of ten mirrors.

Now it is possible to conceive of a situation in which he can see the box *only* with the aid of ten mirrors, that is, no "direct observation" of the box is possible. Since the image of the box in the tenth mirror may be poor, the observer might doubt that the box exists. If he could instead see the box with the aid of nine mirrors instead of ten, the image would be clearer and he would have less doubt that the box exists. There would be even less doubt with eight mirrors, etc. With but one mirror, there would be very little doubt. A key question in this discussion and in the whole science-Christian faith discussion is this: When the last mirror is removed, is the last vestige of doubt concerning the existence of the box also removed?

The removal of the final mirror does *not* remove all doubt. With the use of mirrors, the observer utilizes certain laws of geometric optics — natural laws — to deduce that the box exists. *After the last mirror is removed, the observer continues to use laws of geometric optics.* There is nothing absolute about "direct" vision. The laws of geometric optics which the observer uses in "direct" vision are laws he learned early in his life. Similarly, any observation could be shown to depend ultimately upon laws formulated when the senses are first used. These laws are, of course, primitive. Yet, they are the end product of the process involving first perception and then generalization or law formulation. Thus, the observations which lead to the formulation of the electrical attraction law, for example, are themselves laws based on that which is more fundamental, namely, uninterpreted perceptions.

The more barriers that separate the observer and the observed — in the example these barriers are mirrors — the greater is the use of man-made laws. With greater use of these laws, the possibility of incorrect deduction increases. One might be fairly certain when no mirror is used that the box is indeed on the table. A water molecule cannot be seen "directly"; how certain can one be of its structure, i.e., the arrangement of the atoms in the molecule? In principle, the problem of molecular structure is no different from the problem of the box on the table. The different level of certainty in the two cases arises only because one must use *many* man-made laws in the one case and only a *few* in the other.

The difference does not arise because there is no need for man-made laws in the problem of the box on the table.

Thus, in the strict sense the scientist does not work with observations and deductions. Instead, he works with less complicated and more complicated observations, with the more complicated observations susceptible to error, and with all observations involving an element of deduction.

In spite of these considerations, it is convenient to speak of observed facts and natural laws. *What is to be guarded against is resting an argument on any supposed qualitative difference between observed facts and natural laws.*

C. *Quantitative Aspects of Natural Law.* The law of electrical attraction includes more than the qualitative idea that there is attraction between negative and positive particles. The law also states how *much* attraction exists. For example, if there is a certain amount of attraction between two point charges at a certain distance, the law states that there is only one-fourth as much attraction if the distance between the charges is doubled. The complete statements of natural laws always include such information.

In order to include in the expression of a law this quantitative information, the law is often expressed using mathematical symbols. A word description of a law is a mathematical description, but it is usually more convenient to use mathematical symbols. When a law is expressed in terms of mathematical symbols, mathematical operations – such as operations using algebra or calculus – on the expression of the law can be carried out. These operations can, in principle, be carried out on the word description of a law, but such a procedure would be forbiddingly cumbersome. Natural laws are rendered immeasurably more useful when these mathematical operations are carried out on the symbolic forms of the laws. For example, an entire branch of physics, classical mechanics, has been developed by performing mathematical operations on fewer than five natural laws. (The exact number depends upon how the laws are stated.)

Thus, the use of mathematics is very desirable in a study of natural law. There are even instances in which a law cannot be given in any form other than the mathematical one. With these laws there is no comprehensible physical picture. Usually, however, mathematics is only a tool in the hand of the scientist. It is a much-used tool, and consequently the impression is often given that mathematics is an essential part of natural law. Assuming that mathematics is essential to natural law tends to confuse the issue.

D. *Changes in Natural Law.* While there are both qualitative and quantitative aspects of natural laws, there is normally more certainty concerning the qualitative. Thus, one could be more certain of the fact of electrical attraction between particles than the amount of that attraction.

It is meaningless to speak of a negative electron-positive proton pair in which there is no attraction. Is it meaningless to speak of such a pair in which the *amount* of attraction is different, even though the distance between the particles is the same? Attraction is, as shown above, a direct consequence of the nature of these particles. A careful examination of the nature of the particles reveals that the amount of attraction is just as much a consequence of the nature of these particles. Neither the qualitative nor the quantitative aspects of natural laws can be separated from the nature of whatever obeys these laws.

Thus, if there ever was either a qualitative or a quantitative change in a natural law, there was also a change in the nature of whatever obeyed the law. This conclusion is given because it has frequently been suggested that natural laws have changed. Such a change, whether abrupt or continual, cannot be separated from a change in the nature of that which obeys the law, whether it is a particle, light, or anything else.

The question of whether or not natural law has changed is frequently raised in connection with a consideration of what occurred during and immediately after the six days of Genesis 1. If there were changes in natural law, certain present-day observations concerning events which took place in the past (for example, astronomical and geological events) cannot be interpreted using currently-valid natural laws. The change-in-natural-law suggestion has usually been made to refute the idea that the universe is very old.

Some of the particular questions concerning the time of creation are discussed in later chapters. One conclusion can, however, be given now. Natural law and that which obeys the law cannot be separated. Therefore, a cessation of creation (at the end of the sixth day) cannot mean that the creation of matter (and whatever else was created) ended and that some part of natural law changed at a later time. The behavior of what was created depends upon its nature, and one cannot separate behavior and nature. In the discussion of origin in a later chapter, the meaning of "cessation of creation" is discussed. Whatever cessation means, therefore, it cannot mean that creation ceased but that natural law later changed.

Since the electron and proton particles have already been created, it might seem that laws such as the law of electrical attraction cannot be changing now. However, concluding that a given law is not changing is not warranted. If observations led physicists to conclude that the quantitative aspect of the law of electrical attraction is changing, a new law, incorporating the change-with-time factor, could then be formulated. The presently-accepted law would thus be a special case of the new, more general law, and the new law would be a more fundamental natural law than the presently-accepted law. If the new law of electrical attraction, incorporating the time factor, were correct, then the *most fundamental* law of electrical attraction is unchanging. If, however, this new formulation were discovered to be a special case of still another law, then this other more general law would be the unchanging law. In other words, the most fundamental natural law scientists formulate is always an unchanging law.

Thus, if a presently-accepted natural law is shown to hold only for the present, it is not thereby proved that God is still creating. Natural law and the nature of that which obeys the law are indeed bound up together. God's creative acts involved the most fundamental law and the corresponding nature of what obeys the law. There is no need to hold that creation itself continues.

It is therefore possible that some presently-held natural laws, and whatever obeys those laws, are not the same as the corresponding laws and particles which existed during and immediately after the six creation days. If there has been a change, this change either is or is not susceptible to scientific inquiry. If it is susceptible, there is no reason to accept the change unless it has been proven scientifically. It was shown earlier that the term "natural law" does not refer to that which cannot proceed from observation. Therefore, if a hypothesized change in natural law is in the realm of science, it cannot be accepted unless it meets the usual criteria related to observability.

If a change in natural law is not susceptible to scientific inquiry, the change can only be a miracle. In addition, a postulated miraculous change in natural law implies a permanent change, and therefore a change in creation itself is involved. Consequently, any such postulated miraculous change, a change which includes a change in both nature and behavior, must be harmonized with the Biblical concept of the cessation of creation at the end of the sixth day of creation. Thus, it is not enough merely to assume that there has been a change in natural law.

The conditions for accepting a suggested change in natural law are therefore stringent for the Bible-believer who understands the man-centeredness of the natural law concept. It is not surprising that there have been only a very few reasonable suggestions that the presently-accepted natural laws have not always held.

Miracles

It has often been stated that miracles represent interventions of God in his natural law. Sometimes God is said to "suspend" his law for the purpose of performing a miracle. Another suggestion is that miracles are special concentrations of God's guidance of human affairs.[2] In these suggestions there is the idea that God's hand is present in miraculous events in a way that it is not present in other events. These suggestions bear investigation.

In the previous section it was shown that there is no clear line of distinction between a fact and a law. Furthermore, laws are not isolated: they are related to each other, and many laws are but special cases of more general laws. There seems to be an unbroken network of facts and laws. Man, however, does not perceive all the connections in the network. While the idea of the network is derived from an examination of man's natural laws, the network depends upon God's law, not man's.

A typical miraculous fact to be fitted into the network of facts and laws is the floating on water of the ax-head of a son of an Old Testament prophet (II Kings 6:6). In another miracle an angel appeared to Balaam's donkey and the donkey spoke (Num. 22:27-28). The virgin birth of Jesus was a miracle (Matt. 1:18). To understand how these and other Biblical miracles fit into the fact-law network, God's purpose in performing them must first be understood. This purpose is discussed next, and after that it is suggested how miracles fit into the fact-law network.

A. *The Purpose of Biblical Miracles.* Every miracle impresses upon us the infinite power of God. When we learn of the miracles just cited or of any of the many others recorded in the Bible, we learn that God operates in a dimension we do not understand. His great power consists not only in the magnitude of his acts, as demonstrated in, for example, his ability to cause or to stop great storms. The power which miracles reveal to us is also a *kind* of power we do not have. There is no human power of the *kind* God uses when he raises one from the dead.

As a man becomes aware of God's infinite power, through miracles, he begins to believe that God can perform other miracles as well. He realizes that God is not limited to man's generalizations of events, man's natural laws. The Bible presents miracles in a manner which indicates that the basic purpose of a miracle is to convince the beholder that God can perform not only this miracle but also the miracle of redemption. In this connection, it is instructive to consider three Biblical examples.

(1) Miracles did not accompany the preaching of Peter and Paul for the purpose of attracting a crowd to see this curiosity. Peter and Paul said in their preaching that God can save sinners; if they were correct, *they were saying that God can perform a miracle.* To prove to their

hearers that the God of whom they spoke could save them, they performed miracles in his name.

Peter's preaching could have ended with his imprisonment which is recorded in Acts 12:4. Instead, God caused him to be released. God could have used non-miraculous means, but instead he used a miracle, the appearance of an angel, to be the means whereby Peter was released. Therefore, everyone who later heard Peter preach or who read his epistles, written after this imprisonment, could know that the word which Peter brought was brought only because of a miracle God had performed.

(2) Moses' word to the Israelites also received miraculous verification. When Moses asked God how the people would know that he spoke the truth, God gave him the power to perform miracles. Moses was able to change his rod into a serpent, to make his hand leprous, and to change water to blood (Ex. 4:1-9). Moses could thus prove that he represented the God who had infinite power, enough power to free them from the Egyptians.

(3) John explained the purpose of the miracles he recorded:

> And many other signs truly did Jesus in the presence of his disciples, which are not written in this book: but these are written, that ye might believe that Jesus is the Christ, the Son of God; and that believing ye might have life through his name. (John 20:30-31)

For John, it was enough to link Jesus with the ultimate source of power, God, by means of miracles. By performing miracles, Jesus proved that he was himself that ultimate source of power. Finally, because he was that source of power, he would give life to whoever read and believed John's message.

In summary, in the passages cited and in other Biblical passages describing miracles, it is shown that the Biblical miracle demonstrates that the God who performs the miracle possesses all power.

B. *The "Natural" and the "Supernatural" Mixed.* Sinful man does not deduce from non-miraculous events that he can depend completely upon God. There is no natural law that teaches us that we can depend completely upon God, even though an important reason for which natural laws are formulated is to enable us to predict what will occur in future situations in which we find ourselves. If, however, we do believe that God can take care of us, we believe that God operates above natural law, which, after all, is only man's law.

For the Israelites in Egypt, and later in Israel, God's care included the idea that he would supply their daily needs. Rain or manna would fall as needed; their enemies would be confounded, whether by drought, pestilence, or some other means. To help the Israelites, angels would sometimes appear, perhaps to kill the enemies of God's people. There would be heavenly visions, such as those of Isaiah. There would be healing. There would be the ultimate miracle, the Incarnation.

Such care was outside the realm of natural law. Natural law does not, for example, predict a correlation between observance of God's commandments and the regularity of rainfall. Natural law, since it is comprised of man's generalizations, based on non-miraculous events, cannot tell us when angels will appear. Notice how unexpected is the conclusion which can now be made. When men saw angels, miraculous healings, or something else which was miraculous, they were using the same eyes, ears, and other senses they used when they observed "natural" events. Those with Paul on the road to Damascus heard the *sound* of a voice (Acts 9:7). When God said, "Thou art my beloved Son; in thee I am well pleased" (Luke 3:22), there was *sound.* These sounds were heard with ears that also heard "natural" sounds. Likewise, beholders of miracles saw miraculous light with eyes which could see "natural" light also. With miracles there is an intermixing of the "natural" and the "supernatural." Miracles are windows onto a whole segment of creation, the so-called "supernatural" world, concerning which natural law can tell us nothing.

The concept of natural law which defines the limits of the natural world is man-centered in that natural law can reach no farther than man can comprehend. Therefore, the division between "natural" and "supernatural" is defined by man's limitations. But, man is fallible, and he does not know precisely what his limitations are. Therefore, from man's point of view the line between the "natural" and the "supernatural" is uncertain.

Since there are physical consequences of miracles, miracles cannot be separated from the fact-law network. Both miraculous events and non-miraculous events have given rise to sensations of sight and sound, to the satisfying of hunger, to the feeling of well-being which results from healing, and to many other physical consequences. There are consequently two important conclusions to be made. First, God must be the cause of all the events in the network of fact and law, since there cannot be two ultimate causes of inter-related events and since God is undeniably the cause of miracles. Second, since man cannot ultimately explain miracles, neither can he ultimately explain non-miraculous events. Consider what the situation would be if man *could* ultimately explain non-miraculous events. Ultimate explanation implies that all the causes of a given event are understood. However, in the chain of causes of some non-miraculous events, there are "miracles" which are not explained. Therefore, man can never formulate a law so complete that it would describe the functioning of the universe as if it were a giant machine.

Other arguments, which do not depend upon the existence of Biblical miracles, also show that no such ultimate law is possible. (Cf. Chapter 5.)

In all this the Christian, who believes the Bible, has less of a problem than does the natural man. The Christian knows that there is a law, not the natural law of man, which links together miraculous and non-miraculous events. The Christian knows that the connections in the fact-law network are all there, even though he does not understand many of these connections, and his inability to understand these connections is not of primary importance to him. The natural man, however, does have a problem with this question. Miracles do not fit into his scheme, and so he must reject them. For the natural man, miracles are non-facts.

C. *Miracles and God's Law.* The universally-true law, the ultimate law, is the law which only God knows. Laws such as the law of gravity, the law of electrical attraction, and the more sophisticated unified field theory are not true in *all* of God's creation. God could conceivably have given us in the Bible universally valid physical laws, but he did not do so. Had he given us in the Bible the law of gravitation as we state it, we would not expect to find in the Bible the miracle of the ax-head floating on the water. The law of gravitation would then be an inviolable law of God, not a formulation of man.

God's universally-true law has perfect predictive properties. Scientifically, "predictive" refers to the description of what will occur when certain conditions are given. With respect to the Bible, "predictive" refers to an element of prophecy, a description of future events. Starting from present conditions, a perfect law will predict accurately both future conditions and the consequences of these conditions. Thus, the Biblical and scientific meanings of "predictive" coincide when it is a perfect, universal law that predicts. The presence of so much correct prediction in the Bible is one more way in which there is seen God's complete control of all events, according to a law not understood by man.

The ideas concerning miracles proposed above can lead to difficulty if care is not taken. When some of these ideas are applied by Lever[3] they seem to lead to an untenable position. Lever quotes extensively from Diemer; Diemer in turn claims that the idea of one law for all of creation was held by Augustine, but that Aquinas held to the idea of an "ordinary" nature into which God intervenes for the purpose of performing miracles. Aquinas' idea has largely prevailed, and consequently there now is among both Roman Catholics and Protestants the scholastic natural-supernatural distinction.

Lever, who accepts a God-directed evolutionary scheme, uses the one-law idea in a questionable manner. He concludes that the existence of but one law suggests there were no sudden creations of animals and plants during the six creation days. Any such creation, says Lever, would have been an intervention of God into his creation.

It seems that Lever does not make maximum use of the one-law idea. If, for example, the miracle of Moses' rod becoming a serpent (Ex. 4:2-3) is accepted, and taken to be an event consistent with God's one law, the instantaneous creation-from-nothing of serpents during the six creation days can also be accepted. The rod was not a necessary or even a significant precursor of the serpent. Similarly, the Bible does not indicate that the manna given to the Israelites in the wilderness was anything other than manna created from nothing (Ex. 16:14-15). Christians, who accept miracles, do not have a problem in so interpreting the appearance of manna. If the sudden appearance of manna is in harmony with God's one law, so is the earlier sudden appearance of plants used for other food also in harmony with that one law. Jesus' miraculous provision of a large amount of food using only a few loaves and fish (Matt. 14:15-21, 15:32-38) is another example of this point. These miracles are essentially creations-from-nothing. If these miracles are accepted, then an earlier sudden creation of fish and the components of bread is equally in harmony with the idea that there is but one law.

J. Daane expressed disagreement with the one-law idea, as given by the author.[4] Daane is opposed to treating all facts equally. He says that God's hand is not associated with the facts of an atom in the same way his hand is associated with Christ's healing a leper. The question of Daane's view of facts and their relation to the cross of Christ (a question discussed at length by others) seems not to be at issue here. God is certainly active in both the facts associated with the atom and in the healing of the leper. How God relates these activities is another question. Perhaps man will never understand this relation. God is the cause of all events, and *in this sense* all facts are to be treated equally. The one-law idea does not tell us which facts are ultimately the most important. The one-law idea does, however, express the man-centeredness of natural law, and the God-centeredness of the unity in creation.

REFERENCES

1. van der Laan, H., *A Christian Appreciation of Physical Science,* Guardian Pub. Co., Hamilton, Ontario, 1966; p. 42.
2. van Riessen, H., *The Christian Approach to Science,* Guardian Pub. Co., Hamilton, Ontario, 1960; p. 59.
3. Lever, J., *Creation and Evolution* (tr. P. Berkhout), Grand Rapids International Publications, Grand Rapids, Mich., 1958; pp. 209 ff.
4. The article, "On Miracles" by R. Maatman, *Torch and Trumpet, 4,* 8 (Aug.-Sept., 1954) was answered by J. Daane in "An Inherited Epistemology: I," *Calvin Forum, 20,* 186 (1955).

5 *The Natural Man's View of the Universe*

While the terms "the Christian approach" and "the natural man's approach" are idealized concepts – since a person tends to be neither a consistent Christian nor a consistent natural man – there is a basic difference between the Christian who possesses some of the traits of the natural man and the non-Christian.

The Christian, even though he may harbor many incorrect ideas about the universe, desires to receive all his answers from God. There is a God-created harmony between all men and the rest of the created universe, but the natural man never attempts to receive from God ultimate answers to the profound questions concerning the universe. He views the Christian approach as simplistic and a deterrent to further research. The natural man asks many questions concerning the universe, but, unlike the Christian, he does not find a unifying answer. He finds many answers, answers which assume, but which are not derived from, the idea that there is no God.

When the Christian states, "God made all things," his subsequent work consists of implementing this statement, a statement which contains within it ultimate answers to the profound questions concerning the universe. For the natural man, the denial of God is the assumption upon which he builds. His procedure leads to many questions, because he can use nothing outside of the physical universe.

We shall now discuss some of the problems which arise from considering the universe to be a closed system.

Some of the Natural Man's Problems Concerning the Universe

A. *Accepting and Rejecting Evidence.* Because the natural man accepts the existence only of that which he can find within the physical universe, there exists for him no "supernatural." He rejects the possibility that there exist angels, demons, or anything else not describable by natural law. This rejection is not merely the consequence of the offense which the natural man takes to the Biblical message. He rejects the miraculous in all non-Biblical religions as well.

The natural man believes that his acceptance or rejection of evidence as he considers scientific problems, involves his use of reason, a careful evaluation of evidence, and an arriving at a conclusion as objectively as is humanly possible; and he believes that he has an open mind towards new evidence. But does the natural man use this procedure when the evidence for miracles is presented?

If an event thought to be miraculous can be explained by natural causes, then the natural man will accept the possibility that the event occurred. For example, if a fish is found which is large enough to swallow a man, the story of Jonah can be explained naturally and the natural man will admit that the event could have occurred.

If, however, no natural explanation can be found the natural man uses another method. He attempts to discredit the evidence. The girl living in Judea who reports that she bore a child without an earthly father is called a liar and a fornicator. In some instances the natural man attempts to show that ancient texts have been altered, and that there has been an injection of the miraculous into the text in some romantic manner. For example, the miracles associated with Balaam's donkey are said to have been legends incorporated along with other stories into the Numbers text.

The natural man thus becomes involved in circular reasoning. He rejects the possibility of miracles — without evaluating the evidence — because to accept miracles is to assume there is a force outside the universe. He rejects the possibility of a force outside the universe because, he says, there is no evidence for such a force. Thus he has decided to reject beforehand the very evidence which could show him that there is a force outside the universe, and that miracles are possible. The natural man has made his mind the arbiter of whether or not there are miracles. In this way, it is seen that the natural man has no good criterion for accepting and rejecting evidence. The line drawn between acceptable and unacceptable evidence is thus entirely the product of whim, of an arbitrary decision.

B. *Finite or Infinite Age of the Universe?* The Christian learns from the Bible that there was a beginning. The Christian knew long before the era of modern science that the answer to this scientific question is in the Bible. (The Christian position on this question is discussed in Chapter 8.) What is to be noted here is that this is one more instance in which it is incorrect to maintain that the Bible does not speak to scientific questions.

The natural man attempts to decide, using his self-imposed rules, whether or not the universe had a beginning in time. This problem is extremely difficult for him, since each of the possible solutions has unpleasant consequences for his position.

The concept of natural law always implies that if certain conditions are given, there are certain consequences: a cause-and-effect relationship is assumed. There seems to be no set of conditions which could have as its consequence the *ultimate beginning*. Therefore, an ultimate beginning seems to lie outside the realm of natural law. The natural man is committed to the idea that the universe is comprchensible, and therefore he is becoming more and more certain that there never was a beginning, that the universe existed from eternity. The position is summarized by Schlegel:

> In view of the theoretical difficulties associated with the models that give an infinite past, the evolutionary model, with a creation event that brought it into being, may be appealing to some, particularly to those who for reasons of a traditional religious belief would like to find the universe to have been formed in an event at some time in the finite past. We have emphasized, however, that for the scientist a postulated creation-of-the-universe is generally an arbitrary event, unrelated to other natural processes, and not amenable to investigation by the usual theoretical or observational tools of science. Also, philosophically, a single creation event poses enormous metaphysical questions (even, no doubt, as it does solve some problems). On the matter of creation, as on others, scientists should avoid dogmatism; a scientifically unexplained creation, however, must itself be supported by evidence and theory before it can be accepted into the structure of science.[1]

Several different possible histories of the universe have been discussed by scientists. Galaxies are evidently moving apart. (An example of a galaxy is the Milky Way, of which we are a part.) Consequently, in the "big bang" cosmology (i.e., science of the universe) it is postulated that at one time, billions of years ago, the universe was very small; perhaps it existed at only a point. This has been a popular theory among cosmologists. However, this theory implies there was a beginning and therefore the natural man is becoming more and more certain that this theory does not provide the ultimate answer. Consequently, an extension of the "big bang" theory has been suggested. According to the extended theory, prior to the explosion, there was an implosion, etc., so that there have been an infinite number of explosion-implosion cycles. In these cycles, the minimum size of the universe is much larger than the small region previously considered. There are physical problems with this oscillating-universe model, and efforts to solve these problems are being made.

Another cosmology is the "steady state" model of the universe. It is recognized that galaxies are moving apart and it is postulated that within any large region of the universe new galaxies are forming from matter which is being created, as other galaxies are leaving the region. The term "steady state" is used because any large region remains the same as it both gains and loses galaxies. The entire process is said to have been continuing for an infinite length of time.

Since the natural man is increasingly unwilling to accept the idea of a single creation event billions of years ago, the idea of creation in the steady-state theory also presents a problem to the natural man, although his attitude towards the steady-state problem is not nearly as critical as his attitude towards the idea of a single creation event. In the quotation cited above, Schlegel expressed his attitude towards a single creation event; compare with this his attitude towards steady-state creation:

> The spontaneous appearance of matter [which the steady state] theory re-

> quires has disturbed many physicists and astronomers; for, such a production is an outright violation of the long-established principle of conservation of mass-energy. Still, the actual amount of matter-production required is very small; about one hydrogen atom in a billion years for every liter [approximately a quart] of space will suffice to maintain a steady galactic density. . . .
> . . . The study of creation [in the steady-state theory] may . . . be pursued in the same way as is that of any other natural process; and, it would be the hope of the steady-state theorist that eventually the process of the appearance of matter in space would be elucidated, just as have many other natural processes.[2]

Thus, for Schlegel a creation event billions of years ago must ". . . be supported by evidence and theory *before* it can be accepted into the structure of science." (Italics added.) The study of creation in the steady-state theory, however, may ". . . be pursued in the same way as is that of any other natural process"; it is quite legitimate for the steady-state theorist to hope that ". . . *eventually* the process of the appearance of matter in space would be elucidated." (Italics added.) A creation event of long ago is not to be accepted *until* it can be proved scientifically; continuous creation in the steady-state theory can be a working hypothesis, with the hope that the hypothesis will eventually be scientifically proved and understood. A scientist can be a steady-state theorist and a respectable scientist; however, no such status would be accorded one who claimed to be a single-creation-event theorist.

The natural man who accepts steady-state creation, even if only as a working hypothesis, is driven into an uncomfortable position. He attempts to construct a model of the universe based on natural law, law he can formulate using observations he makes. He proceeds assuming the universe is comprehensible, i.e., it can, in principle, be described by man's natural laws. Now, there are two kinds of natural law: a law may be a summary of observation, such as the law of electrical attraction; or a law may be accepted because its opposite cannot be shown to be true. An example of this second kind of law is the law of conservation of mass-energy, which states that mass-energy is neither created nor destroyed. Many chemical and nuclear reactions (nuclear reactions are reactions which often involve radioactivity) were observed carefully, and no contradiction to the mass-energy conservation idea could be found. On the basis of the absence of contradictory evidence, it was decided that mass-energy can be neither created nor destroyed. If there is indeed steady-state creation of matter, then the mass-energy conservation law is not universally true. However, the amount of creation postulated in steady-state theory is too small to be observed, and the mass-energy conservation law, which makes claims only for the observable, cannot be used to disprove steady-state creation.

The problem of the natural man who tentatively accepts steady-state creation lies in the logical difficulty he has in replacing the mass-energy conservation law with steady-state creation, a new law. The mass-energy conservation law was accepted *only* for observable systems. If this law was found to hold, within experimental error, in one hundred chemical reactions, the general statement of the law implied that mass-energy is conserved, within the same experimental error, in an as-yet-unstudied one-hundred-and-first chemical reactions. But if steady-state creation is to supplant the mass-energy conservation law, *the new law should also be based on observations which do not contradict the new law.* These observations must be made of systems in which the "law" could *conceivably* be deduced to be false if it is indeed false. Thus, when observations of chemical reactions were made, it would have been possible experimentally to observe, for example, a decrease in the mass-energy of the reacting system if the mass-energy was in fact destroyed during reaction.

There have been no observations whereby the *creation* part of steady-state creation theory can be found to be true or false. Those who accept steady-state creation in any way, whether as a law or as a working hypothesis, are in the position of a natural man who might have accepted mass-energy conservation before any system was observed to exhibit such conservation. It has, of course, been suggested that certain astronomical observations indicate something about the truth or falsity of creation in the steady-state theory. None of these suggestions compares with the yes-or-no results which were obtained for the chemical and nuclear reacting systems, results which could have contradicted the mass-energy conservation law being tested. There are other suggestions concerning steady-state creation observations which may be made in the future. What remains, however, is that some persons who accept the position of the natural man tentatively accept steady-state creation, while they reject a single creation event because it has not been scientifically proved. They accept, even if only tentatively, a new law without subjecting the proposed law to the same kind of test they demand of other natural laws. In other words, they propose a new "natural" law which is not based on nature. According to the ground rules of the natural man, such a "law" cannot be a law.

Some cosmological evidence favors certain aspects of steady-state theory, although such evidence does not indicate that there is continual creation. Other evidence recently obtained suggests that the universe may not be uniform; if so, the steady-state creation theory as originally proposed is not correct. Consequently, modifications of the original theory have been suggested.

In any event, postulating a universe without a beginning always presents a problem to the natural man, regardless of whether he accepts the steady-state theory, the oscillating universe theory, or any other such theory. Man does not pretend to understand the concept of an absence of a beginning. Incomprehensibility is not a problem for those who do not attempt to use the comprehensibility starting point in formulating a view of the universe. However, the no-beginning idea is presented by the natural man just because he claims his view of the universe must be comprehensible. If he did not claim comprehensibility, he could logically accept a single creation event. Therefore, the natural man is led by his insistence on comprehensibility to postulate that which is incomprehensible, the no-beginning concept. Here also the natural man's own ground rules have driven him into an impossible position. If the Christian were convinced that the Bible taught him there was no beginning, he could conceivably accept this incomprehensible idea because he is not committed to comprehensibility. The natural man does not have this option.

C. *Finite or Infinite Size of the Universe?* The Bible teaches that it is God alone who is infinite. The Bible does not teach that the universe is infinite. The Bible speaks in many ways of God's creation as less than God himself. The clay is never equal to the potter; the universe is finite. The Bible therefore provided the answer to this important scientific question, and here also the Christian can depend upon the Bible to teach him correct science.

For the natural man, the question of whether or not the universe is finite has not been settled. In the steady-state creation theory it is assumed that expansion has been continuing for an infinite time, and that therefore the universe is infinite. In the other cosmologies there is also the possibility of an infinite universe. For example, even if the natural man accepts the big-bang theory, he has no assurance that there was only one big bang: there could be many, or an infinite number, of universes such as ours. There is at present much discussion among cosmologists concerning whether or not the universe is finite. This discussion includes both interpretations of presently-available data and also suggestions concerning future observations which might aid in making a decision.

The natural man does not have here merely a scientific problem which the Christian does not have. In this problem the natural man has the same kind of internal inconsistency in his system which he has when he considers whether or not the universe is infinitely old. For him, the only authority is the authority of natural science, since natural science rests on human comprehensibility. The natural man is willing to entertain the idea of an infinite universe, even though he cannot comprehend such a concept. Thus, he admits into his thinking here also the incomprehensible, and he violates his own rules. He cannot protest that the concept may *some day* be comprehensible, unless he is willing to reject the idea until that time. But *rejecting* the idea of an infinite universe until the universe is proved to be infinite is the one thing he will not do.

D. *Cause and Effect.* There is another fundamental matter which puzzles the natural man, but which does not present a problem to the Christian. The three questions discussed above have been discussed in one form or another for centuries, but there has been an unexpected turn of events concerning cause-and-effect in the twentieth century. These events have been unsettling for the natural man; as a consequence, there has been a crisis in his thinking.

The crucial question is, "Is this a universe in which man can verify the cause-and-effect relationship?" Up until the twentieth century, it would have been said that if enough is known about a system, its future behavior can be predicted exactly. For example, if two billiard balls collide, we can, it was thought, predict exactly what will happen after collision if their masses, their initial positions, speeds, directions, etc., are known. Certain conditions would always be followed by certain consequences. There is, it was thought, a cause-and-effect law operative in the world of natural science. But when quantum theory, including the "uncertainty principle," was developed, it became evident that *a cause-and-effect statement cannot be made for very small particles,* e.g., electrons.

Strictly speaking, a cause-and-effect law cannot be postulated for billiard balls either, since they are but aggregates of the very small particles whose behavior is uncertain. The *average* behavior of the aggregates can, however, be predicted, if there is included in "predicted" the concept of a very small amount of inherent uncertainty. Even assuming there is no human error, the exact position of the two billiard balls after collision can never be predicted, although the amount of uncertainty is infinitesimal. Regardless of the size of the theoretical uncertainty, one fact remains: on the most fundamental level, the level of the smallest particles, the particles which are the component parts of any masses which our senses perceive, the validity of a cause-and-effect relationship between events cannot be tested.

Notice how earth-shaking this non-testability is for the modern scientist. D'Abro writes:

> The new quantum theory indicates that we cannot, even in theory, test the classical doctrine of causality. On the strength of these findings, the quantum theorists have rejected classical causality, adopting the principle of complementarity [a principle which states one can omit certain other vital elements in the description of a system if the cause-and-effect relationship is to be kept] in its place.

> Those who are opposed to the revolutionary implications of the principle will presumably claim that the impossibility (even theoretical) of testing a doctrine does not necessarily prove that the doctrine is erroneous; and on this basis they might propose to retain the classical philosophy. But the weakness of this claim is that *if we accept doctrines which cannot be verified in principle, no restriction is placed on caprice; anything may then be postulated with impunity.*[3] (Italics added.)

According to D'Abro, if a cause-and-effect relationship between events cannot be *tested,* it cannot be maintained that such a relationship *exists.* Probably no answer to a fundamental question is more disturbing to the natural man. His goal has been comprehension of the universe. He has assumed that the universe is comprehensible, that man can, in principle, understand in the finest detail any part of it. His assumption of comprehensibility is just that which gives rise to his problem. He cannot claim that one fundamental event causes another, for then ". . . anything may then be postulated with impunity." The natural man is thus forced to reject the law of causal relation between events involving the most fundamental particles. But this is the one law he can never afford to reject.

This dilemma is very serious, and consequently many scientists who accept the fundamental assumption of the natural man have expended a considerable amount of effort to remove the difficulty. First, as explained by D'Abro, one can keep the cause-and-effect law if he is willing to give up some vital element of the description of the system. Such an approach has some importance, but it does not solve the problem of the natural man. He cannot sacrifice part of the system's description in order to comprehend *all* of the universe. This approach shifts the problem: there is still an element of incompleteness. His position remains vulnerable.

Second, the cause-and-effect law can be kept if it is applied only on a statistical basis. Use of this principle will enable one to predict, as mentioned above, the behavior of the billiard ball with only infinitesimal error. Similarly, if a single atom receives energy, it will store this energy and emit it as a photon (a packet of light energy) at some future time. It cannot be predicted, however, *when* the photon will be emitted. Yet, if a large number of atoms receives energy, reliable statements about *average* behavior can be made, even as an insurance company can predict mortality rates but not the time of death of a given individual. Since the scientist can make such statistical predictions about atoms, scientific work is possible. Thus, he can make useful predictions about colliding billiard balls, and the extremely small amount of uncertainty there is in these predictions is of no importance in practical situations.

The usefulness of the statistical approach tends to mask the problem which made this approach necessary. The problem of uncertainty, of incomplete comprehensibility on the most fundamental level still exists. If it were discovered that a single-particle cause-and-effect law *could* be proved, the effort to substitute a statistical cause-and-effect law would be abandoned immediately with a sigh of relief. But such a single-particle law is lacking. Even as the princess of the fairy tale could feel the pea through twenty blankets, the scientist knows that masking the original problem does not remove it. He can still feel that pea.

Another indication that the statistical approach cannot be the sought-for answer is that there are systems consisting of so few particles that the statistical approach is meaningless; but these are, nevertheless, systems which can be studied experimentally. Modern instruments can actually, in favorable cases, study the behavior of single atoms. The behavior of the atomic nucleus in a nuclear reaction, or the behavior of an electron of the atom involved in the emission of a photon, cannot be treated statistically when only one atom is involved.

Third, the cause-and-effect dilemma is sometimes attacked by maintaining that the problem is not a real one, and that further scientific work will show this. Einstein, for one, felt that the description used (the "quantum mechanical" description) was incomplete, and that eventually the element of certainty would be re-introduced into all physical situations. This approach is appealing to many, especially non-scientists, who often view science as changing so rapidly that no "it cannot be done" statements are to be allowed. Yet, the weight of the evidence against this approach is very great. One would be putting his faith in an unproven assumption were he to accept this approach. As of now, there is no suitable cause-and-effect law. The natural man, if he would be consistent, ought to be very dissatisfied with accepting this approach, an approach in which he accepts a "solution" only because he wants it to be the solution. The natural man has in other contexts ruled out-of-bounds the idea of accepting unproven statements.

There is always the possibility, in spite of the weight of scientific evidence, that the cause-and-effect dilemma may be removed. There still remains one serious objection to the position of the natural man relative to this question. This objection also holds if there is a removal of any of the other three problems posed, the problems the natural man has with the criteria for evidence and the finitude of the universe with respect to time and size. Because this objection to the position of the natural man is general, it is discussed separately in the next section, entitled, "The Importance of These Problems."

The Christian cannot scorn the development of modern physical theory, theory which includes these uncertainty ideas. The achievements in this field are far more magnificent than the average non-scientist realizes. Some of the twentieth-century investigators in fundamental physics are probably among the most brilliant men who have ever lived. Therefore, the answer of the Christian to the cause-and-effect question is not a criticism of any scientific work. The Christian does, however, criticize the comprehensibility assumption of the natural man. In natural science, this assumption means that the sum of the natural law which man can deduce accurately describes the universe. The Christian says that there is not a one-to-one correspondence between man's correlations, natural law, and the actual law which governs the universe. There does exist a cause-and-effect law which is universally, objectively, and absolutely true. This law the Christian does discuss. The law which does govern the universe is the law which only God knows. The universe is not completely comprehensible to man. It is completely comprehensible to God alone. But the universe is not only comprehended by God; it is also governed by him. The Bible indicates in many ways that God governs the universe. Some typical passages in which this is taught are the following:

> . . . Upholding all things by the word of his power. (Heb. 1:3)
>
> . . . His kingdom ruleth over all. (Ps. 103:19)
>
> Are not two sparrows sold for a farthing? and one of them shall not fall on the ground without your Father. But the very hairs of your head are all numbered. (Matt. 10:29-30)

In Job 38-41 God tells Job his power extends to the bottom of the sea, to the animal world, to the stars, and to many other places. While the natural man is confronted with the annoying cause-and-effect questions, the Christian's response is that of Job:

> I know that thou canst do every thing, and that no thought can be withholden from thee. (Job 42:2)

The natural man charges the Christian with adhering to a position which frustrates scientific investigation. This frustration supposedly arises because the Christian introduces a non-natural element, the sovereignty of God, into the cause-and-effect question. There is no point, says the natural man, in discussing cause-and-effect if a transcendent God provides all causes. However, the Christian is not frustrated in scientific investigation. He knows that scientific investigation is meaningful because it is investigation of the creation which God governs; in a feeble way, he is able to think God's thoughts after him. Rather, it is the basic assumption of the natural man, the assumption that man can comprehend the universe, which is frustrating. His assumption means that reality is just that which man can comprehend. This idea leads to a dead end. For him, nothing outside of our universe is real, since nothing else is comprehensible. Yet, he cannot find a satisfying cause-and-effect law *within* our universe. Since such a law cannot be found within our universe, he who claims that only this universe exists cannot claim the universe to be comprehensible.

At this dead end, a new concept must be introduced. This is the concept that the universe is not self-contained, that there is a force outside of the universe which acts upon it. It is not suggested, of course, that God acts in the universe only by acting in the realm which is inaccessible to man. God's sovereignty is in no way limited. How God foreordains, with or without the human uncertainty concept, and how his foreordination and human free will can co-exist, is a mystery. Man cannot resolve this problem. Nor can man, by showing that our universe is not self-contained, deduce the existence of the living, triune God. Man can, however, show that the god of ultimate human comprehensibility does not exist.

The Importance of These Problems

The problems raised in the previous section are perhaps more important than one might first suppose. One might expect that further scientific work would solve the problems, and that the dilemmas are only apparent, not real. Is it not possible that the natural man will solve some of these problems by using his own criteria?

It is probable that at least some of these four problems cannot even in principle be solved using the natural man's criteria. Suppose, however, for the sake of the argument, that these four problems can and will be solved. Suppose that the natural man convinces himself he has developed a truly objective criterion which enables him to rule out evidence of miracles; that he has solved to his own satisfaction the problems concerning the finitude of the universe with respect to both time and space; that he has removed the cause-and-effect problem. Even then, the natural man would not have removed an objection he himself would have. The natural man cannot ever be certain he has solved any one of these problems correctly. It should be noted, in addition, that the natural man now doubts, concerning at least one of these problems – the cause-and-effect problem – that it is in principle solvable; and, even if he later thinks that he has solved this problem, he will still remember that he once doubted it is solvable. Therefore, he will always be uncertain of any solution he develops.

Even though an astronomer would develop evidence that convinces all cosmologists that the universe is of finite size, there would always be the possibility that later evidence would change the conclusion. The scientific conclusion which lives because of the scientific approach can also die because of the scientific approach. It should not be otherwise. The only *absolute* knowledge we can have is that which God reveals to us. Other knowledge is relative and tentative.

The four questions discussed in this chapter have one element in common. They are questions which in one way or another are concerned with the limits of science. These questions can be re-phrased, emphasizing the limiting concept: How is one to draw the limiting line which separates valid evidence from invalid "evidence"? Is the present age of the universe limited? Is the size of the universe limited? Is the starting point (a kind of limit) of scientific work within the universe or outside of it?

Thus, the natural man has difficulty with the "edges" of science. He fares well in scientific investigation when he works away from the edges. For example, when he formulates natural laws, he can do well if he does not maintain that these laws rule out miracles. If he does not claim his laws reveal something about whether or not there was a beginning, or a finite size, his laws function well. If he carries out scientific work without insisting that he understand ultimate cause, he can obtain fruitful results. It is as if the natural man correctly works out the geography of part of the earth's surface, erring only when he supposes that Atlas is on the other side of the earth, holding it so that it does not fall.

But it is just at these outer limits that the natural man is least likely to surrender. His approach is that he, with his comprehensibility assumption, is best suited to solve these "outer limits" problems. If man is to be the measure of all things, then the natural man would say, it is *only* man who can answer these ultimate questions concerning the universe.

Consequently, the Christian and the natural man do not have common ground, insofar as "common ground" refers to the foundation, the assumptions, upon which each builds. The basic assumptions of the Christian and of the natural man — those assumptions referred to in connection with the four "outer limits" problems — are not compatible. Frequently, "common ground" is used to denote common ideas which spring from different fundamental assumptions. Thus, even though the Christian and the natural man start out poles apart, they agree perfectly on, for example, the periodic classification of the elements, a concept of extreme importance in chemistry. This kind of "common ground" is limited, and to illustrate how limited it is, an analogy may be helpful.

Consider the experiences of a certain pair of parachutists who jump simultaneously from two planes flying close together. The parachutists are able to converse as they fall. The one parachutist opens his chute with confidence because he trusts the person who packed the chute. The other parachutist has some misgivings as he attempts to open his chute: he remembers that prior to jumping he did not trust the person who had packed his chute, and consequently the parachutist re-packed the chute, even though re-packing is against the rules. He is an amateur at packing chutes, and so his chute does not open, and he is destroyed. His friend, with whom he communicated during the first part of their descent, descends safely.

As the parachutists were falling before the one chute opened, they seemed to be in equal situations. One could conceivably hold that they then were on common ground. They could communicate with each other. Yet, the one parachutist had done something as he prepared for the descent which insured that he would be killed. The starting points of the two were thus quite different, and consequently the ends of their experiences would also be radically different. Notice that the brief time that they were in apparently equal situations is not separated from the beginning and the end of their experiences. It is not as if their temporarily-equal situations concerned a matter wholly separated from their descents. The time they were in apparently equal situations was for each parachutist a *necessary link* between his starting point and his fate.

When the natural man and the Christian carry out scientific work away from the "edges" of science, they seem to be in equal situations. Yet, like the parachutists, their starting points and the ultimate consequences of their starting points are quite different. What is most important is that their apparently equal situations are *necessary links* between their starting points and where they are going.

Here is the crux of the matter: the day-to-day work of the scientist is the *necessary link* between his starting point and the ultimate destiny of both the scientist and his work. The existence of this link can be seen for both the natural man and the Christian.

The natural man begins in science with the idea that man can, in principle, become the master of his environment. He can comprehend it and control it. The natural man expects that eventually he will fully understand nature and control his environment so that he can live the "good life." Some disease has been virtually eradicated; he expects that science will achieve much more. Some areas are now affluent, even though they once knew poverty because they seemed to be subject to the vagaries of nature; the natural man expects this improvement process to continue indefinitely, so that all areas of the world will improve. Where there was once a small life expectancy and little leisure, there is now not only longer life and more leisure, but also many more ways to use leisure time enjoyably; again, the natural man expects science to achieve more and more. In other words, on the theoretical level the goal the natural man has set for himself is complete understanding; on the practical level, he aims to be master by using science.

For the natural man, the god-concept is the concept of an unknown force accounting for events man cannot otherwise understand. The natural man expects that a primitive society would believe lightning and thunder to be evidence that there is a god. The natural man says that as man understands lightning and thunder in terms of natural forces, man no longer needs a god. Mastering theory means there is no need to postulate the existence of a god. Similarly, the practical victories of science in matters of comfort, health, etc., make it less and less necessary to rely upon a god for comfort, for health, or for any other thing man desires.

There is therefore no part of the science of the natural man, whether it is engineering or quantum theory, the study of distant galaxies or unraveling the structure of the molecules found in living cells, which is not a necessary link between his man-honoring, God-dishonoring starting point, and his personal goal as well as his goal for science.

In spite of all this, the Christian can work with the natural man. All the scientific areas of activity — engineering, medicine, astronomy, etc. — are also for the Christian necessary links between his starting point and his goal for himself and for his science. The Christian is impressed with physical theory (even though it is a theory posited by man and subject to revision) because it reveals the inner structure of God's creation. This theory shows that God is the God of harmony, and that the universe is not cluttered with isolated facts. There is a subtle harmony which God allows men to see. When in some branch these theoretical ideas are developed enough so that man is given health, comfort or some other desirable thing, the Christian sees a providential God. His creation is a good creation, with great potential. Furthermore, the Christian sees in these desirable things a faint foretaste of that which God has promised him in eternal life in heaven. The God who promises and provides the good things of this life also promises infinitely more in a heavenly life.

No man works without motivation. The labor of each day in the life of the scientist is labelled in one of two ways, and the label indicates the kind of motivation the scientist has. The labels read either, "Performed by man in his attempt to become the master of the universe," or, "Performed by man in obedience to God for the purpose of honoring God."

REFERENCES

1. Schlegel, R., *Completeness in Science,* Appleton-Century-Crofts, New York, N. Y.,
2. *Ibid.,* pp. 123-124.
3. D'Abro, A., *The Rise of the New Physics,* Vol. II, Dover Publications, 1951; p. 955.

6 In Defense of Science

Man can receive fantastic benefits from science, even though there are limitations on the science of the natural man. These benefits can be received regardless of the world-and-life views of the scientists who are responsible for these benefits.

Within the body of science, apart from the limits discussed earlier, correct results can be obtained by both the natural man and the Christian. However, many people consider large areas of science – sometimes even all of science – quite suspect. It is therefore necessary to defend scientific work in the face of the accusations which are frequently made against it. The particular accusations often made by Christians are considered here.

Accusations Against Science

A. *Science and Chance.* It is said, "The science of the natural man rests on nothing but chance." Ultimately, this statement is true, but it is true in only a very subtle way, as was shown in the last chapter. However, the quoted statement can convey a wrong meaning. Several years ago a university class was told by its instructor that the uncertainty principle (concerning the cause-and-effect question) means that one cannot measure *accurately* both the position and velocity of a race car. They were impressed, until they learned in a later class period that the error would be very, very much less than the accuracy of the instruments used to measure position and velocity. Such indeterminacy is meaningful when small particles, such as electrons, are considered. Statistically, the behavior of large aggregates of small particles, such as racing cars, can be predicted with great (but not perfect) accuracy.

It is the statistical description which provides the predictions, usually as accurately as desired; and here is the real triumph of science. Here is found "truth" as science knows truth. Here there are examples in which mathematics is used to describe correctly the outcome of a scientific experiment before the experiment is performed. For example, in thermodynamics (a branch of physics) the values of the "translational entropies" of the inert gases are more accurately calculated mathematically than can be determined experimentally. The statement that science rests on nothing but chance seems to take into account neither this example, nor the many others like it.

Because the science of statistics is reliable, there is virtual certainty that if a coin is flipped one million times, the number of heads will be between 49% and 51%. When the behavior of grams or ounces of molecules is considered, the degree of certainty becomes almost infinitely greater. How is it then possible to maintain that God is active in the behavior of the coin and the molecules?

Man cannot answer this question any more than he can answer the question, "How can God predestine even while man is responsible for his actions?" a question which has exercised theologians for centuries. There is a parallel between these two questions. It may be helpful for those who accept the Christian faith, but who maintain that it is not possible that God predestines while man is responsible, to consider this parallel. If we must choose between divine predestination and human responsibility, we ought also to choose *between* divine control of aggregates of atoms and human statistical prediction of the behavior of these aggregates. Are not the two problems related, so that if the one problem is beyond our understanding, so also is the other?

If the idea of divine control of aggregates of atoms is relinquished, the hand of God is neither in our affairs nor in any other event in the universe. The Christian faith would then be a farce. On the other hand, if the idea of the statistical prediction of the behavior of aggregates of particles is relinquished, the validity of science and of even the most elementary human experiences is denied. Consistency would then demand that one cannot expect the proverbial apple to fall to the ground.

There is no Christian alternative to accepting *both* the divine control of aggregates of atoms and the human statistical prediction of the behavior of aggregates. The relationship between these two is a mystery. Yet, the Christian must accept these two ideas. In much the same way neither predestination nor human responsibility can be relinquished.

B. *Changing Science.* It is said, "Scientific knowledge is constantly in flux, and therefore scientists should not speak glibly of 'scientific truth.'"

Scientific ideas do not change nearly as much as is popularly supposed. In fact, it is rather uncommon in well-established scientific disciplines for ideas to be supplanted by new ones. Such a sweeping claim must be explained.

If current knowledge in a particular scientific discipline is represented by the area within a circle, then current investigation is being carried out only at the circumference of a circle. As investigation adds to scientific knowledge, the circle becomes larger and the discipline becomes better established. The knowledge within the original circle will very likely not be affected by the addition. New knowledge rarely proves older ideas to be wrong.

What *does* happen is that older ideas may be seen in a new context. Consider the example of quantum mechanics and Newton's laws. Some persons have erred in claiming that modern quantum mechanics and the uncertainty principle have shown Newton's laws of motion, formulated in the seventeenth century, to be incorrect. Newton's laws are incorrect for the smallest particles, particles which were unknown in the seventeenth century, but those laws are correct (within experimental error) for the larger objects for which the laws were formulated. While it surprised many scientists to learn from quantum mechanics that a satisfactory cause-and-effect law for the smallest particles could not be derived, it is not fair to say that Newton was incorrect in his frame of reference.

There seem to be several special reasons the average non-scientist believes that all of science is changing. First, he usually learns of scientific developments through some form of communication which tends to popularize these developments. Newspapers, magazines, and news broadcasts tend to emphasize the *newness* of a discovery. There is usually little need to show any relationship to older ideas. Only if a research worker has a startling new idea which contradicts older ideas, is the relationship between the new and the old shown. Very frequently the news presentation favors the one who seems to throw over the old. Often such a new idea is eventually shown to be untenable. The rejection of the new to return to the old is not glamorous and very likely such a rejection is not of interest to the purveyor of news.

Were one to obtain his ideas about scientific progress from the popular news media, he would be led to believe that the scientific research front is chaotic indeed. Many non-scientists may not be aware of the nature of the scientific press, the means scientists use to communicate with each other. The scientific press consists largely of journals, usually difficult to read. Here the rash claims of the newspaper or magazine article are not seen. Dull sobriety is the rule. Yet, here is "where the action is." A slightly new insight here, a variation on an old experiment there – these are the pieces which, when added together, make the explosion of scientific knowledge.

In the scientific press there are differences of opinion, but only a few times in this century have old, universally-accepted ideas been overthrown. Frequently, tentatively-held ideas are dropped. However, the abandonment of such ideas is not what critics refer to when they maintain that science is in such a state of flux that scientific pronouncements must be taken with a grain of salt.

It is not only the glamorizing of new developments and opinions that tends to present a distorted picture of science. Over-emphasizing scientific hoaxes is equally undesirable. One would certainly expect to find some dishonest opportunists among scientists. Yet, there have been only a very few scientific hoaxes. Scientists are generally honest reporters. They are not honest in these matters because science makes men honest, although there certainly are those who teach this foolish idea. For a scientist, cheating in reporting is not the "in" thing. Such cheating can be detected, and the perpetrator is trusted about as much as a physician would be were he to be found guilty of administering poison instead of medicine. Hoaxes (and conspiracies, which are compound hoaxes, and which are discussed in a later chapter) may be ruled out as a meaningful factor in modern science.

Another reason the non-scientist tends to believe that there is an essential instability in scientific knowledge, is the revolutionary impact of science on our way of life. The non-scientist knows, for example, what happened in the transportation revolution: the animal as a means of land transportation was partially supplanted by the railroad; there followed, in order, the automobile, the propellor plane, and the jet plane. He knows that each new method of transportation was considered impossible until its practicality was demonstrated. Each brought about a transportation revolution. Each changed the old order irrevocably. Each is a product of science. Does this not, many persons reason, show that science never speaks the last word on a subject? In fact, are not the thousands of inventions reported each year proof of the same point?

Equating invention with a change in science is to misunderstand the nature of invention. An invention rarely, if ever, represents the application of a newly-discovered natural law. The discovery of a new natural law rarely means that an older idea is invalid. Instead, inventions usually prove that certain existing ideas are correct. This could be demonstrated by considering the series of inventions which made automobiles, planes, etc., possible. Instead, an extreme example is chosen.

The series of inventions which has made space travel possible has depended not only upon fundamental knowledge concerning thrust, the nature of fuels, etc., but also upon general knowledge of the solar system. The sizes, densities, gravitational fields, and distances from the earth to the moon, the planets, and the sun are but some of the facts which were deduced before man explored space with manned and unmanned space probes. Man's ability to probe space has depended upon the correctness of many earlier scientific ideas. Thus, the complicated sequential reasoning used to deduce much of the nature of Venus, Mars and the moon, including the sizes and shapes of their orbits and the relation of their orbits to the earth's orbit, has been verified by certain of the successful space probes made in recent years.

What is popularly meant by "progress" in science is actually the production of new inventions. An invention is a synthesis of old, correct ideas. There is even now a very large amount of scientific progress which could be achieved if there were correct combinations of presently-known natural laws. For example, new "wonder drugs" have been made and will be made not because old "laws" are found not to be laws, or because new methods of synthesis are discovered, but because the right compound is chosen, out of a myriad of possibilities, and it is synthesized by the appropriate combination of known methods.

What is the scientific truth ("truth" only as it is used in science) upon which scientific progress is based? Examples of scientific truth are the three laws of thermodynamics (e.g., mass-energy is conserved in natural processes), the law of gravitation, the law of electrical attraction, Newton's three laws of motion, and the laws underlying quantum mechanics. But there is no absolute distinction between a law, which is a human correlation, and the observations which enable the scientist to make the correlations (cf. Ch. 4). Scientific truth thus consists not only of laws, but also of observations. Consider the example of light. Its behavior has been observed, and the observations themselves are part of scientific truth. Some of these observations have been generalized in order that laws may be formulated. One such law is that light is reflected a certain way at a plane surface. Other observations of the behavior of light can be used to determine that light travels 186,000 miles per second in a vacuum. Observations and summaries of observations such as these constitute the body of scientific knowledge.

While certain subjects are well understood, there are gaps in our knowledge which we know to be gaps. We can fill some of these gaps without the development of new techniques or laws. Merely by continued observation much can be learned, for example, about the chemistry of the cell nucleus. We could know much more about the earth, particularly its interior, if there were geological observations made at large, but not impossible, depths. By continued work using present methods and techniques chemists can obtain much more information concerning molecular and crystal structure. Present techniques and methods can be used to obtain more information about the nucleus of the atom. These are only a few examples of what can be obtained with presently-available methods and techniques. With *new* techniques of investigation which probably will be developed, the horizon is infinitely broader.

In addition, products and processes of practical application very likely will be developed because of the basic knowledge we now have. Transistors, wonder drugs, giant computer "brains," and the various synthetic fibers are all examples of revolutionary products which were new only in that they were new combinations of already-existing concepts and materials. There is no reason to doubt that such inventions will continue to appear. In all these realizations of potentialities, whether it is the elucidation of galactic structure or the development of a new paint which can be used on concrete, scientific knowledge increases not by refuting the old, but by using existing knowledge as a foundation.

C. *Differing Scientific Opinions.* It is said, "Scientists aren't as correct as they claim to be. There are other points of view. Sometimes scientists disagree among themselves." The question of controversy among scientists is often overemphasized. Concerning the day-to-day progress involving small steps, there is continual discussion and some disagreement. Virtually no new idea goes unchallenged. However, there are large areas of universal agreement.

In spite of this agreement, pamphlets, booklets, and books which claim to overturn much of science abound. The writer of such material often claims there is a scientific conspiracy to prevent dissemination of his ideas. These "works" are written polemically; and, in general, they lack even the outward marks of scholarship. Similar approaches are used by certain individuals and groups active in political or religious debates. In science and in these other areas this kind of writing can usually be ignored. It is, however, necessary to mention such material in this book because some of it is concerned with the relation between science and the Christian faith. Those writings are characterized by (1) the claim that many scientists are conspiring to suppress an idea, or (2) the unscholarly nature of the work.

Concerning the first characteristic, there is in the present state of affairs in science no warrant for the conspiracy claim. In the history of science there have been individuals who were prejudiced against a certain work and who consequently attempted, sometimes successfully, to prevent its publication. There are surely individuals active today who are attempting the same thing. But individual action does not constitute a scientific conspiracy. Reasonable ideas can and do receive a hearing. There are many journals in each of the scientific disciplines, and it simply is not correct to claim that they are all controlled by conspirators. In one important and emotional scientific question, the question of biological evolution, the viewpoint of the minority *is* voiced in the places where opinions are formed, that is, in the articles and books which scientists read. This minority view has been expressed in responsible journals and books, as shown in Chapter 13.

In spite of the opportunity the worker has to present his views to his colleagues in the journals, books, etc., which his colleagues read, complaints are continually voiced. For instance, Morris has said,

> One reason for the apparent dearth of anti-evolutionary sentiment is that the major publishing houses and periodicals are completely and exclusively under the control of leaders who are evolutionists. If anyone questions this, let him try to get a serious scientific article or book published refuting evolution! Or even a Letter to the Editor! The only outlet for such literature seems to be through conservative or private media.[1]

Nor is it only the anti-evolutionists who make such a complaint.

Putnam, who claimed that Negroes are inferior to whites and that Negroes could by an evolutionary process surpass whites,[2] stated that the race-equality idea rests on a "pseudo-scientific hoax," and that some scientists secretly hold the same opinion:

> . . . I found professional scientists aplenty who saw what I saw. And I discovered something else. One prize-winning Northern scientist whom I visited at his home in a Northern city asked me, after I had been seated a few minutes in his living room, whether I was sure I had not been followed. Another disclosed in the privacy of his study that he had evidence he was being checked by mulattoes at his lectures. All, when first approached, were hesitant, withdrawn and fearful, and the reason was not far to seek. Their employers on whom their livelihood depended — the universities, the museums, the foundations — were either controlled by equalitarians or were intimidated by the race taboo. The scientists whom these institutions employed, if they were ever to hint at the truth, must do so deviously, under wraps over wraps, half seeming to say the opposite.[3]

Concerning the second characteristic which condemns a work, it is self-evident that an article or a book which purports to be scientific must be scholarly. While it is difficult to define "scholarly," there are certain obvious marks of scholarship. One important mark of scholarship is that the work be addressed to scholars. It is improper for one to present a revolutionary scientific idea to the public before the new idea has been subjected to the give-and-take of scientific discussion. Therefore, one test of whether or not a scientific idea presented to the public should be taken seriously, is whether or not the idea has actually been discussed in the scientific literature. There is absolutely no reason for the scientific worker to avoid this demanding, but constructive, step.

Another mark of scholarship which should be insisted upon is the requirement that the writer who presents a new idea be an expert in the field about which he writes. (Naturally, this strict requirement applies only to the originator of an idea, not to one who writes about or teaches the ideas of others.) Virtually all the far-out literature here being condemned has been produced by those who have little reason to know much of the fields they have "investigated." Both scientists and non-scientists have spoken freely on subjects they understand only superficially.

A scientist is not able to speak with authority on all scientific subjects. To speak with authority, he must be trained *in that particular discipline*. It is as reasonable to accept the biological "discoveries" of a chemist, as it is to accept the medical care given by a mechanical engineer. If one wishes to learn about the deeper meaning of "race," he should consult the anthropologist, not the politician. If he wishes to learn about the methods of dating rocks by the use of radioactivity decay schemes, he should give heed to the man who has worked on such matters, not to one who has not. The pseudo-scientific literature is cluttered with the highly-opinionated views of non-experts. We understand when the farmer laughs at the city-bred man who tells the farmer how to farm. We ought also to understand the scientist who is unwilling to give attention to one who speaks outside his own area of competency.

The scientist confronted with the statements of one working outside his own area of competency has a sense of frustration. Anyone can understand this frustration if he imagines an outsider telling *him* how to carry out *his* work. Today much of the popular press ascribes unusual power to the novice. Legends concerning scientific amateurs abound. One can easily see the flaw in this approach if he tries to suppose that beginners in *his* kind of work can surpass all the experts.

D. *Science Versus the Bible.* It is said, "If the Bible is accepted literally, much of science must be rejected. No amount of scientific argument can refute that fact."

Suppose the unscholarly work already referred to is not taken seriously. Suppose further that the current scientific work of competent Christians and non-Christians is examined. Then a conflict between science and the Bible does not arise.

There are, of course, many non-Christian scientists who have come to conclusions entirely in conflict with the Bible. Some of these conclusions were described in Chapter 5, where the natural man's view of the universe was discussed; and other similar conclusions are described later in this book. What is very significant in all this is that there is a disagreement *among the scientists themselves* on just these conclusions which run counter to the Bible. These conclusions are *not* in that body of knowledge universally held and virtually fixed. For example, concerning the finitude of the size or age of the universe, or concerning biological evolution, there are many scientists, Christian and non-Christian, in the relevant disciplines who disagree with the anti-Biblical position. When there is a difference of opinion among scientists, it is not fair to cite only one opinion in an attempt to prove science and the Bible are in conflict.

There is no guarantee that there will always be significant scientific dissent against the scientific conclusions which contradict the Bible. According to the Bible, many new weapons will be used against the Christian in the last days. At that time a unanimity among scientists concerning some un-Biblical concept could well be used against Christians. However, Christians do not arm themselves properly against that day when they falsely appraise the present situation.

The Christian who finds absolute truth only in the Bible is not in the difficult position in which one finds himself if he accepts absolute authority from another (sometimes, additional) source. No doubt many Roman Catholic scientists have encountered a conflict as they attempt to believe in the "real presence" of Christ in the bread and wine of the sacrament. Without question, this doctrine derived from tradition was formulated in a day when the approach to the understanding of matter, such as bread and wine, was mystical, as contrasted with the present analytical approach. (Interestingly, the Bible uses neither of these approaches.)

It is appropriate for the Christian to thank God for his providential prevention of a conflict between the Bible and the matters upon which scientists agree. It is improper for the Christian to condemn the present situation, a situation in which God's restraining hand can be clearly seen.

E. *Christians Allowed to Carry Out Scientific Work.* It is said, "Almost all scientists are non-Christians, and the non-Christian stifles the work of Christians. In some disciplines it is virtually impossible for Christians to obtain advanced degrees."

There is little evidence that the percentage of scientists who are Christian is different from the percentage in other professions. The day may come when the Christian is not permitted to serve on university faculties and in similar positions of trust. That day has arrived in Iron Curtain countries, but it is not fair or correct to state that it has arrived in the West.

In the United States it has often been claimed that a biologist who does not accept evolution cannot teach in a state university. This is not true, since anti-evolutionary biologists are teaching on such faculties. It has often been claimed that a student opposed to evolution cannot obtain an advanced biology degree. Without doubt some faithful Christians have been badly treated in this matter. Satan has no desire to be tolerant. Yet, there is adequate proof that others have been treated fairly. It is not that the unbeliever has such a desire to be tolerant. Rather, it is God who restrains the unbeliever. Consequently, a sweeping condemnation of present-day university faculties is not justified. Such condemnation should be treated by the Christian just as idle gossip is to be treated. Rather than dealing in unfair generalities, the Christian should be thankful that God has kept for himself, in the world of the scientist, many who have not bent the knee to Baal.

REFERENCES

1. Morris, H. M., *The Twilight of Evolution,* Baker Book House, Grand Rapids, Mich., 1963; p. 27.
2. Putnam, C., *Race and Reason, a Yankee View,* Public Affairs Press, Washington D. C., 1961; p. 53.
3. *Ibid.,* p. 19.

7 Man's Study of Nature

There are several reasons man should engage in scientific activity, activity in which he studies nature. It is of interest and importance at this point to focus attention on the Christian reasons for engaging in scientific activity. Christians generally agree as to what these reasons are, although they do not necessarily state them in the same way.

Christian Reasons for Scientific Activity

A. *God Commands Scientific Activity.* God gave several instructions, "creation ordinances," to Adam and Eve before they sinned, and these ordinances are to hold for all time. They are to be distinguished from the laws which have special meaning in a world that also contains sin. Therefore, neither sin nor its consequences are to be associated with God's purpose in giving these ordinances. For example, the relationship between husband and wife, which was ordained before the fall into sin, is not to be considered necessary only because sin is in the world. Thus, the creation ordinances shed light on the underlying structure of a perfect creation. One creation ordinance is of special interest here:

> And God blessed them, and God said unto them, Be fruitful, and multiply, and replenish the earth, and subdue it: and have dominion over the fish of the sea, and over the fowl of the air, and over every living thing that moveth upon the earth. (Gen. 1:28)

Apparently the complex command given here was to be carried out in different ways in different times. While the command seems to instruct men in all ages to develop agriculture, it was possible only after man had acquired certain knowledge that the concept of subduing the earth included the development of mining. Similarly, there was a time when it became appropriate to include the working of metal and the making of tents. (See Gen. 4:20-22.)

Replenishing the earth and subduing it in our day includes also the command to carry out scientific work. Because of modern scientific discoveries, Genesis 1:28 has a deep meaning for those who read the Bible with confidence in its perfection. On the other hand, the Christian who carries out scientific work *without* believing that the Bible speaks authoritatively in science seems to have a weak reason for carrying out such work. If the Bible does not anticipate modern science in Philippians 2:10, the passage in which there is according to some persons an unscientific reference to a three-story universe (see Chapter 3), consistency seems to demand that neither is science anticipated in Genesis 1:28.

The most important reason for carrying out any activity is that God has commanded it. The Christian can discern at least two additional reasons which indicate that scientific activity is included in the command to subdue the earth.

B. *Man Is to Use Creation.* Scientific progress can make life more pleasant for man. Scientific exploitation of what God has created does not in itself provide the "good life," for no man has a good life, with peace in his heart, unless he has peace with God through Christ. But God does provide for his creation through science. Adam was told by God that as a consequence of his sin his work would be unpleasant (Gen. 3:17-19). By giving us modern science, God has lightened that punishment. Death (Gen. 3:3) and pain in childbirth (Gen. 3:16) came because of sin. In giving us science, God has permitted the lengthening of life and the diminishing of pain in childbirth.

In addition, many products of science are useful in bringing the gospel to those who would not otherwise hear it. The revolution in human life brought about by modern communication and transportation has been described many times. It is interesting to observe what has happened in Bible-science interaction. An important factor in bringing about the modern explosion of scientific knowledge was the widespread dissemination of the Bible. Modern science has been primarily responsible for the unbelievable, almost alarming, growth in world population. Without the communication-transportation revolution made possible by modern science, large numbers of those who have been brought into the Christian faith never would have heard the gospel.

Perhaps none of these ways of using creation would have been necessary had man not sinned. It was stated earlier that the command to subdue the earth was given in a sinless-world context. The following shows that science does more than overcome some of the effects of sin.

C. *Man Marvels at God's Works.* To understand that marvelling at God's works is an important reason for studying science, consider the experience of a mountain climber.

A mountain climber can suddenly come upon an awesome panorama and feel very humble. He may not be sure why he has such a feeling. He may feel he is in the presence of a kind of perfection. Or, he may feel his own striving in life is quite meaningless when compared with the scene which seems not only to be vast but also complete. With both of these emotions it is his heart, his inner person, which comes into personal contact with a part of creation. He will experience one or both of these emotions whether or not he thinks deeply about life and the meaning of life experiences.

Before the modern era the most arresting man-nature interactions were usually thought of in terms of phenomena the ordinary man could understand or observe – a mountain view, the raging ocean, the starry sky. Even then it was realized that not all men could have all these memorable experiences. Only a few were strong enough to climb mountains; not all could live on or near oceans.

Today the experiences of the scientist parallel the experience of adventurers of an earlier day. But horizons are now much more magnificent. Some galaxies now being studied are billions of light years away. The familiar body organs are now known to be unbelievably complex. The atom is known to be a myriad of particles and/or fields, so complex that it may be impossible to find truly "fundamental" particles. The more deeply involved in these new discoveries the scientist becomes, the greater is the impact creation makes upon him. His experience is similar to that of the mountain climber in two ways. First, relatively few have such experiences. Second, the impact occurs *whether or not* he thinks about and analyzes life's experiences.

The adventurer of years gone by may have thought deeply about man's relation to creation. Today God has enabled the Christian scientist to see mountain-tops not visible to earlier explorers. The Christian scientist receives great comfort from his reflections. One of the sources of this comfort is the knowledge that his Savior is involved in creation; John says, speaking of Christ,

> All things were made by him; and without him was not any thing made that was made. (John 1:3)

Reflective scientists who are not Christian and who accept the position of the natural man usually admit that they do not experience such comfort. They seek meaning, but they do not find it.

Any attempt to add science to the Christian faith, as if these two can exist in separate compartments in a man, cannot succeed. There are no separate compartments in the Christian man. Some scientists see God's wonders in nature before they know him through the Bible. Some accept Christ through the Bible before they become scientists. Regardless of which comes first in a man's life, Christian faith or scientific knowledge, the two become bound together; and as he grows spiritually the Christian sees relations in the whole context in which God has placed him. With this growth, the Christian scientist understands more of the creation command to subdue the earth. As the scientist who is a Christian obeys God's command in his work, God makes his life meaningful and satisfying.

As scientists appreciate what God has created, they become impressed with the fundamentals of science. It is as if the scientist were digging in an attempt to find the ultimate foundation in a building: the longer he digs, the deeper he goes, and more and more of the foundation is revealed. That which characterizes what the scientist finds as he discovers more of the fundamentals is the *order* in nature. The more of the foundation that is revealed, the more order which appears. The Christian is amazed at what God has done. Since this orderly aspect of God's creation is so important, a separate section is now devoted to it.

Order in Creation

Christians are universally impressed with the order they see in nature. Ministers preach about it; Christian teachers emphasize it; Christian books and magazines refer to it often and in many different ways. Probably there are few adult Christians who have not accepted the idea that God has created an orderly universe. Two questions concerning the concept of order are of importance.

A. *What Does "Order in Creation" Mean?* There is an almost universal human desire to understand nature in terms of a small number of simple natural laws. For millennia men have attempted to understand natural phenomena by showing that the phenomena are the consequences of simple principles. During the modern explosion of scientific knowledge there has been an intense effort to explain widely divergent phenomena – for example, gravitational, electrical, and magnetic phenomena – in terms of a simple law, perhaps one described by a single equation.

The possibility that simple natural laws can explain a wide variety of phenomena reflects the order which exists in creation. Note that men ascribe order to nature insofar as humanly-comprehensible laws are capable of describing nature. These humanly-comprehensible laws need not be laws which are presently known. Men are sufficiently convinced of the existence of order even if such laws have not yet been formulated. For example, the structure of the nucleus of the atom is even today imperfectly understood. Yet, a few decades ago physicists did not doubt that there are simple laws, perhaps only one law, which would provide a satisfying picture of the nucleus. Previous evidence of order left the physicist with no doubt that there is order in the atomic nucleus.

An example taken from elementary physics illustrates that the simplest principles can provide adequate explanations for many seemingly-complex phenomena. Long ago men learned that a moving body moves more rapidly when a force is exerted (in the direction of the motion) on that body. In fact, if the force is doubled, the increase in velocity also doubles. This is a natural law. Another natural law that has been accepted is the law that energy can be neither created nor destroyed, unless mass is destroyed or created. *Using only these two natural laws,* a law, or theorem, known as "Bernoulli's Theorem," can be shown to be true. By means of this theorem, it can be shown that a baseball pitcher can cause the ball to curve if he puts the appropriate spin on the ball. The same theorem says that if two boats are near each other and moving in the same direction, they will be drawn together, with the possibility of collision. Bernoulli's Theorem also explains why the wing of an airplane can lift the plane.

Another example, taken from chemical physics, is given to illustrate what the scientist means by "order." Electrons in the atom act, strangely enough, as if they possess some of the properties of waves. That is, there is a mathematical equation which describes the wave motion of a vibrating solid (similar to the equation for a vibrating string, such as a violin string) and the same equation can be used to describe the behavior of the electron in the atom. The equation applied to the electron is called the "Schrodinger equation." In formulating the Schrodinger equation only a small number of natural laws are used. Two of the laws used are the law associating the electrons with waves and the law of electrical attraction (or repulsion). Almost all chemical facts can either be derived now, or in principle are derivable, from the Schrodinger equation.

To illustrate, Schrodinger's equation is used to explain why color is associated with many compounds of copper (examples are blue vitriol and the green scales which form on statues as weathering occurs) and iron (for example, the color of rust). Atoms of copper, iron, and many other elements which can form colored compounds possess so-called "d-orbitals." Electrons are associated with these orbitals, and in the atoms which form the colored compounds here considered, electrons are associated with some, but not all, of the d-orbitals which are present. This some-but-not-all situation gives rise to color, a consequence of the validity of the Schrodinger equation. In other words, rust is colored because of the existence of a few fundamental laws, laws concerning waves, charges, etc. Color arises in completely different materials for just the same reason.

The Schrodinger equation explains the central correlation of chemistry, the periodicity of the elements. (Inside of any chemistry classroom there is displayed a "periodic chart," which is a chart exhibiting this periodicity.) If the elements are arranged according to their atomic number, groups of elements having similar properties are thereby formed.

When the periodic classification principle was first enunciated a century ago, it became evident that certain elements were missing. Because of this principle, the properties of elements which were later discovered were correctly described. This was a significant triumph for the order concept.

Using this concept of order, chemists were able to resolve the puzzle of the elements zirconium and hafnium. Hafnium is much less abundant than zirconium, and the properties of the two are almost the same. Because of this similarity, hafnium was present in zirconium ores, and its presence was not detected. When certain chemical reactions were used to separate zirconium from impurities, the impurity hafnium behaved the same as zirconium and therefore it was not separated from zirconium. The existence of an element such as hafnium was not doubted, since there was a gap in the periodic classification corresponding to the position hafnium took when it was eventually found. Finally, careful work proved that hafnium was indeed present in zirconium ores. Thus, both the existence of the undiscovered element and its similarity are consequences of the validity of the Schrodinger equation, the unifying equation of chemistry.

The amazing relationships such as those which have been mentioned explain many natural phenomena, and those who study these things are astounded. The magnificence cannot be described adequately for those who have not studied these sciences, no more than words can describe a symphony or a painting. Yet, everyone receives at least a glimpse of the interrelatedness of physical phenomena. The non-scientist marvels at the ordered motion of the heavenly bodies; at the "fortuitous" freezing of a body of water from the top downward, thus preserving aquatic life in most regions; and at an endless number of other phenomena. These phenomena may seem to be very complex, but they depend upon a very few simple laws, such as the law of gravitation. Thus the "order" of the non-scientist and the "order" of the scientist are the same, with the "order" of the scientist including, but not limited to, the non-scientist's idea of "order."

It is seen that the desire of man to put all natural phenomena under one logical "roof" is directly connected with the order man finds in nature. To the extent that man satisfies this desire to unify, he finds order in nature. However, it is important to emphasize that "order in creation" is what God has put in creation, and not necessarily that which man observes. Therefore, the relation of God to the order in creation is now to be considered.

B. *Is Order in Creation Arbitrary?* If order is taken in the subjective sense just described, it is possible that God could have made man incapable of knowing order. All that would then be implied by the concept of a disordered universe is that man would be unable to make correlations demonstrating the order of the universe. This would not mean, however, that God's will, the truly unifying law, would not be operative. There could be disorder in the subjective sense in the ordered universe which God has created, but such a concept of a disordered universe would relate only to man's understanding of the universe, and not to the universe itself and God who created the universe.

Order in creation is not arbitrary if order is that which depends upon the absolute law of God. Order in creation can be arbitrary only in the sense that God could have made man incapable of discerning order. Often the order which is discussed is, in reality, this subjective order, even though this distinction is sometimes not made explicit. When one refers to this subjective order in the universe, it is implied that God has made man so that (1) man desires to put natural phenomena under one logical roof, and (2) man is partially able to satisfy this desire. Man, created in the image of God, is thus created *in harmony* with the rest of creation, enabling him to construct natural laws which are still but feeble imitations of parts of the truly unifying law, God's will.

God could have created a universe in which man would be able (at least in principle) to discern a unifying law, or unifying laws, which would explain all natural phenomena. In reality, however, God created a universe in which occurrences such as miracles cannot be comprehended by the law that man formulates. There are also other limitations to man's knowledge (cf. Chapter 5), limitations often expressed in terms of man's finitude. However, God could have created man capable of comprehending the universe which, after all, is finite.

C. *How Does Man Know There Is Order?* There have always been men who knew that God's will is the ultimate law, and so before the age of science the principle of order in the universe was known, though not as well understood by those who did not know God. Even today, the order which is perceived when scientific law is understood, is only an aid in understanding the ultimate in order, God's will. As scientists discover more order in the universe, they amplify that which the Christian always knew in principle. It is as if all knowledge is a vast picture broken into the pieces of a jig-saw puzzle. The Christian knows that the mixed-up assemblage of pieces fits together, and that man is incapable of fitting together all the pieces. The natural man, however, judges that since some pieces have been fitted together, man will be able to fit together the rest.

Since (1) man is created in harmony with the rest of creation, and (2) God revealed to man from the beginning that his will is ultimate and all-encompassing, man gained knowledge of order, first of all, because God revealed that there is order. Is it fair to say that the natural man knows God's will to be ultimate, even though the natural man denies the existence of God? In this book an attempt is made to show that the Bible provides many facts which are relevant in a scientific discussion. The Bible answers this question about the knowledge of the natural man:

> For the invisible things of him from the creation of the world are clearly seen, being understood by the things that are made, even his eternal power and Godhead; so that they are without excuse: because that, when they knew God, they glorified him not as God, neither were thankful; but became vain in their imaginations, and their foolish heart was darkened. (Rom. 1:20-21)

All men know of God and his power, even of the "invisible things" of God. In spite of the knowledge which the natural man had ("they knew God"), he did not glorify God. His heart became darkened.

Using the teaching of this Romans passage, the thesis concerning order here being presented can now be amplified. All men have known in their hearts (the heart is the innermost being of man) that God's will is the ultimate law. God created man with the desire and the ability to formulate some natural laws, which are infinitesimal parts of that ultimate law. Since men have always known that there is such an ultimate law, man's desire to formulate laws did not arise because of some initial successes in formulation. Rather, the desire is a consequence of the knowledge that there is such a law. Man's ability to formulate natural laws exists because man was created in harmony with the rest of the universe. This harmony between man and the rest of the universe cannot be separated from his innate knowledge that there is an ultimate, all-encompassing law. Thus, order in the universe is fundamentally known by all men. Some men understand that God has revealed order to them. The others, who are aware of this order, have darkened hearts and they believe that they see order only because of man's achievements.

The concept of order in the universe often gives rise to an emotional response in men. The realization that there is order strikes a chord in man's heart. It is no wonder the power of God, as reflected by the order in creation, universally impresses men.

In summary, the various reasons for man to study nature are connected. God commanded man to subdue the earth, and so by implication he commanded man to carry out scientific work. God created man and the rest of the universe in harmony, and thus man can perceive an order in creation. The order which man perceives is just that which makes scientific work possible. Man can obey God's command to subdue the earth because man and the rest of the universe are in harmony. What God requires, he made possible by creating as he did. Many scientific products, whether they are new building materials, hybrid corn, or radar are possible only because of this harmony in creation. Behind the command to subdue, the products of science, and man's response as he marvels at the order in creation, is God's care for his creation, which is the heart of the matter.

8 Concerning Evolution

There are several problem areas in the interaction between the Christian faith and natural science. References to some of these areas have been made in previous chapters. In a book in which the theme is the relation between the Christian faith and natural science, it seems appropriate for at least two reasons that at least one of these problem areas be examined in some detail.

First, the treatment given to the problem serves as an example of the use of some of the principles given in earlier chapters. Second, the problem chosen, evolution, is so important that an extended discussion of the relation between science and the Christian faith is incomplete without some evaluation of this problem, even if primary attention were given to another problem area. For these reasons, the remainder of this book (except for the concluding chapter) is a discussion of evolution, creation, and origins.

A. *Why Evolution Is Discussed.* Of the several possible Christian faith-natural science problem areas, why is evolution chosen for discussion?

Evolution is the most widely-debated problem area of the relation between science and the Christian faith. The debate has been intense, often emotional, as shown by the pseudo-scientific literature discussed in Chapter 6. (Some other important areas of debate are the scientific control and manipulation of the human mind, the control of human genetics, and faith healing.)

The questions raised by a discussion of evolution are of extreme importance to the Christian. These questions concern not only evolution itself but also the methods used to investigate this problem. How we use the Bible to study this question influences our use of the Bible as we study other questions apart from the relation between science and the Christian faith. Concerning the evolution question itself, the answers obtained have relevance for the Christian faith, especially insofar as these answers have bearing on the nature of man. Often it is not realized that the nature of man is at issue. Making a mistake in this most important matter will necessarily distort one's ideas concerning the so-called fundamentals of the Christian faith.

B. *Why the Bible Is to Be Considered First.* Throughout the evolution discussion in this book, an attempt is made to discover *first* what the Bible teaches. This procedure should be followed in the study of any question upon which the Bible sheds light. If this is not done, and one uses another source first, he might sin by contradicting what God states in the Bible. For example, to take an extreme case, it would be sinful for one to be a part of a scientific expedition which searches for the dead body of Christ. After it is determined what the Bible teaches, science may then be used, either because the Bible does not speak directly on this matter, or because science can amplify the Bible. Science is not to be used to prove the truth of the Bible, although there are many ways in which today's science beautifully illustrates facts already known to the Bible-believer.

To illustrate the need for examining the Bible, let it be tentatively assumed that there is a difference of opinion among biologists concerning whether or not biological evolution occurred. On this question the Bible must do one of three things: reveal that biological evolution occurred; reveal that it did not occur; or remain silent concerning this question. To make a conclusion about whether or not the Bible answers the question, one must examine the Bible. It is easy to begin a study of the Biblical position on this question, since it has been claimed that certain Biblical passages teach that biological evolution did not occur. If it is maintained that the Bible does not answer the question concerning biological evolution, it should be shown that these Biblical passages are irrelevant. Usually those who maintain the Bible does not answer this question do not analyze specific passages, but they claim that it is not consistent with the character of the Bible to answer this type of question. This seems to be a clear example of putting the mind of man, as he decides what the Bible teaches, above the Bible. To the extent that the mind of man is made the ultimate authority, the position of the natural man is taken.

C. *Science, an Aid to Bible Study.* Extra-Biblical information, including scientific information, can amplify the Bible. Thus, the findings of non-Biblical history have been profitably used in Bible study.

Consider the Biblical statement that Christ was born

> . . . when the fulness of the time was come. (Gal. 4:4)

What does "the fulness of the time" mean? This phrase means that the Christian gospel could spread because there was a universal language, a universal Roman citizenship, and a good system of communication and transportation in the Roman empire. The phrase also means that at the time of Christ's birth there was widespread moral corruption and dissatisfaction with the pagan religions. This information, which makes the Bible much richer for us, is derived from sources outside the Bible.

Another obvious example of help from outside the Bible is the use of ancient literature to understand the grammar of the Bible.

If it is admitted that one kind of extra-Biblical knowledge can aid in understanding the Bible, then it is possible that other kinds of extra-Biblical knowledge can be helpful. It has been objected that while we can know ancient history and ancient grammar very well, our scientific knowledge cannot attain the same level of certainty. This is a mistaken idea, since scientific knowledge and historical knowledge can be known equally well, although neither kind of knowledge is absolute knowledge. Thus, historians can provide information which adds a meaningful context to "the fulness of the time," but historical information which contradicts the idea that Christ came in the fulness of time cannot be regarded as true.

D. *Problems with the Use of Science as an Aid.* Once it is admitted that science can amplify the Bible, it is easy to make mistakes which lead to serious difficulty.

For example, incorrect conclusions have often been made from the following passage:

> But the day of the Lord will come as a thief in the night; in the which the heavens shall pass away with a great noise, and the elements shall melt with fervent heat, the earth also and the works that are therein shall be burned up. (II Peter 3:10)

Some Bible students have suggested that the idea of the elements melting with a fervent heat is a prediction of a giant nuclear holocaust, in which the elements we know could be said to melt. The mistake in such an interpretation lies in the assumption that the modern concept of "element" coincides with the ancient concept. In addition, the text may allow for destruction at a temperature much lower than the temperature achieved in nuclear explosions (since it is only melting, not vaporization, which is mentioned), although the higher nuclear explosion temperatures are not ruled out.

Calvin, who interpreted the Bible so carefully, could also err in applying science to the Bible. He said in his exegesis of Matthew 4:23 that epilepsy is related to the phase of the moon. This idea is a discredited "old wives tale." Calvin also had some peculiar ideas concerning the soul. One can sense that he made improper use of extra-Biblical ideas as he discussed these ideas:

> . . . the organs of the body are directed by the faculties of the soul . . . .
> . . . in sleep [the soul], not only turns and moves itself round, but conceives many useful ideas, reasons on various subjects, and even divines future events.[1]

The nineteenth century Bible commentators Keil and Delitzsch, respected for their careful work, could also make improper use of science in interpreting the Bible. In their commentary on Genesis 1 they state that science shows that the sun and stars are dark and that sunlight "proceeds from an atmosphere which surrounds" the sun. It is difficult to understand why they made such an incredible statement.

In spite of all this, there should not be a rejection of the use of science in interpreting the Bible. Instead, there should be an improvement in the use of science which earlier commentators made. By updating their use of science it is possible that there will result a better understanding of the text.

Men often possess incorrect secular knowledge. Does the use of secular knowledge in understanding the text mean that there is an element of uncertainty in the Christian faith? There is no such uncertainty. Even though the sin-clouded minds of men err, the Holy Spirit guides the Christian's understanding of the Bible, and consequently the redemptive message and the salvation of the Christian is sure. All of the Bible is related to the gospel of salvation, and Christians understand the Bible well enough for their redemption and their living the redeemed life, regardless of whether or not the understanding of a particular passage is aided by extra-Biblical knowledge.

E. *Concerning "Dissection" of the Biblical Text.* In the process of determining the meaning of specific Biblical statements for scientific questions in the evolution discussion which follows, the Bible will of necessity be analyzed carefully even as the scientist analyzes nature. It is then necessary to be very particular about the word-by-word meaning of the text. Without doubt, part of the meaning of some passages cannot be obtained merely by word-by-word analysis, for some passages are clearly allegorical. But whatever "allegorical" means, it does not mean that the precise meaning of each word is of no value. Christ is our example (cf. Chapter 3) for careful use of the individual words of the text.

Many Christians are not pleased with a word-by-word analysis of, for example, Genesis 1. The reason given is that there is a meaning which is seen only when a passage is taken in its entirety. To put it another way, analyzing a flower petal-by-petal does not aid one in perceiving the beauty of the whole flower. There is indeed a value associated with the whole passage which differs from that obtained when the passage is analyzed in detail. However, the word-by-word analysis cannot contradict the meaning obtained from an examination of the passage as a whole. And further, the meaning of the passage as a whole cannot be obtained without a word-by-word analysis.

F. *Disagreement Among Bible-Believers.* Occasionally those who have a high view of the Bible, who believe the Bible to be inerrant, infallible, and, in its entirety, the Word of God, disagree among themselves concerning the evolution question. It is not a happy situation to find it necessary to differ with these defenders of the Bible, for their view of the Bible is just that view which all Christians should take. What is said here on the evolution question is offered in the hope that Bible-believers can become more faithful in their defense of the Bible as holy, infallible, and inerrant. Certain suggestions are made with the hope that the amount of confusion will not increase but decrease, and that the

Bible-believers' defense of the Bible will become more effective. An attempt will be made to take a position which can and will not be successfully attacked by the opponents of the Bible. The natural man is to be given no opportunity to show that the Bible-believers' defense of the Bible is weak. There is a danger that a very strong attack on the Bible may be made if the views of many who have a high view of the Bible do not change.

The position taken here is that the Bible can and does teach the scientist on certain scientific matters. This is also the view of many who believe the Bible but who disagree with some of the conclusions which will be presented. Perhaps those who have a high view of the Bible, who do not accept that popular idea, can help each other. By helping each other, they may in the foreseeable future arrive at the same conclusions.

G. *Dividing the Evolution Question.* The question, "Is evolution correct?" is phrased inadequately. "Evolution" consists of several propositions, and it must be recognized that several evolutionary propositions are being widely discussed today. For convenience, we will make a purely arbitrary division of the evolutionary position into the following elements.

(1) Matter has existed from eternity.

(2) The earth and at least some stars are billions of years old.

(3) Life in the form of one-celled organisms evolved from non-living matter.

(4) All animals and plants evolved from one-celled organisms.

(5) Man's body evolved from animals.

Two observations concerning these propositions must be made. First, these propositions do not stand or fall as a group. For example, one could consistently, although not necessarily correctly, hold that all animals evolved from one-celled organisms ("one-celled organisms" is a phrase which will be used to denote the simplest form of life), thus accepting Proposition 4, and yet deny that these organisms evolved from non-living matter, rejecting Proposition 3. It would even be consistent (again not necessarily correct) to hold that matter has existed from eternity, but that the earth and the stars we know are far less than a billion years old.

Second, these propositions must be considered one at a time. Notice what can happen if this rule is not observed. Some scientists point to the obvious fact of present changes in the earth's crust to prove certain ideas leading to an affirmation of Proposition 2. They then make an illogical jump to Proposition 3. It is one thing to maintain that present mineral compositions indicate there was once a very different kind of atmosphere. It is quite another thing to maintain that these earlier, different conditions, assuming they existed, "allowed" the spontaneous synthesis of living matter.

Some of those who accept none of these five propositions have also confused the various component parts of the evolution concept. For example, proving from the Bible that matter is not eternal, thus refuting Proposition 1, has no bearing on the problem of those biologists whose primary concern is process, and who accept Propositions 4 and 5. Christians should recognize that the argument must eventually include *both* process and ultimate origin, and that answers to *all* of the evolutionary propositions must be included in a truly comprehensive world-and-life view.

In the remainder of the book, these five propositions will be discussed in the order in which they are presented above.

The First Proposition: "Matter Has Existed From Eternity"

In Chapter 5 this proposition was discussed insofar as it concerned a problem the natural man encounters as he considers the universe. The natural man does not know if matter is eternal, whereas the Christian knows from the Bible that it is not eternal. What are the reasons for the Christian position concerning matter?

The Bible states,

> In the beginning God created the heaven and the earth. (Gen. 1:1)

This statement would seem to answer the question to everyone's satisfaction. Yet, there are Christians who hold to the steady-state creation theory or to some modification of it, and who maintain that phrases such as "in the beginning" are to be understood only in the frame of reference of the ordinary persons who have heard and who hear these words. According to the argument, Moses' contemporaries (and also the Bible reader of today) could not comprehend a creation which did not *begin*. It is impossible for man to understand any other situation than one in which all clocks were once set at zero.

This argument seems to limit God in his ability to transmit ideas to men. If matter were eternal, could not God have told us so? Naturally, we cannot understand *how* such a thing is possible, but neither can we understand the idea of creation which implies a beginning. Whether or not man can ultimately understand a Biblical passage is not a criterion that may be used in interpreting any passage of the Bible.

Furthermore, the term "beginning" in Genesis 1:1 must mean "beginning" in the same sense it is used in this discussion, that is, in contrast to the concept of eternal matter. In the creation account which follows Genesis 1:1 there is included the creation of the component parts of the universe the scientist studies. Everything that the scientist studies is described in the creation account, and it is this that God created "in the beginning."

The question about the existence of a beginning is also answered in Psalm 102:

> Of old thou hast laid the foundation of the earth; and the heavens are the work of thy hands. They shall perish, but thou shalt endure: yea, all of them shall wax old like a garment; as a vesture shalt thou change them, and they shall be changed: but thou art the same and thy years shall have no end. (Ps. 102:25-27)

This passage is quoted in Hebrews 1:10-12, where "of old" is given as "in the beginning."

The Psalmist here gives a picture of a God who exists forever, and who existed prior to the time of creation. The concept of a beginning implies for the Psalmist that creation ages; it grows old. There is difficulty in holding to the steady-state theory, or any similar theory, if "all of them shall wax old." The Bible does not present us with a picture of old garments continually being used to make new garments. Such a process could continue for a time, but finally *all* will be old. Even as the earth and the heavens finally end, they also had a beginning. The whole process — beginning, growing old, and ending — is shown to contrast with a changeless God, a God who has neither beginning nor end. The force of the passage would be lost if one were to accept the steady-state theory or a similar theory and claim the passage refers only to the universe as we now know it.

Two Scientific Developments

It might be thought that the falsity of Proposition 1 is so obvious that there is no point in discussing it. After all, one might ask, what besides a beginning is possible? The basic reason for discussing the concept of a beginning is that there are many who maintain that there was no beginning. There are two scientific developments which are responsible for the widespread acceptance of the idea that there was no beginning.

A. *Elaboration of the Evolution Concept.* The scientist who uses the arguments of the natural man usually urges that postulating a beginning precludes further scientific work. For the natural man, only the physical universe is real; laws operating within it are comprehensible by men; and since the concept of a beginning cannot be comprehended by man, no such concept is to be accepted.

This abhorrence of the beginning concept, of the scientifically unexplainable, is exhibited in several places in evolutionary thought. The natural man claims it is not "fruitful" to postulate the sudden appearance of an animal. For him, it is more satisfactory to show that the animal was derived from other living organisms. This reasoning is extended, until it is postulated that all life has been derived from the simplest forms of life. For the same reason, the natural man takes the next step and suggests that living things did not arise suddenly as living things; they evolved from non-living matter. In a similar manner, the sudden appearance of non-living matter is to be avoided. The natural man postulates that matter arises in some way, not presently understood, because of what already exists. The natural man states that our presently-inadequate concepts force us to associate what already exists with time, and that this forced association suggests a beginning. The natural man assumes that human concepts will not always be limited to time. By these steps, the development of evolution since Darwin has led to the postulation of no beginning, of an eternally-existing universe.

B. *The Downfall of Intuition.* The scientific development which made it possible for the idea of no beginning to gain some scientific respectability was the so-called downfall of intuition, which occurred over the past century. For example, it had been believed – that is, accepted intuitively – that an adequate cause-and-effect law has meaning regardless of the size of the particle. It was pointed out in Chapter 5 that this most "obvious" belief is not correct. Several other ideas associated with atomic physics, such as the simultaneous wave-likeness and particle-likeness of electrons, other particles, and electromagnetic radiation, are also contrary to our intuition. Similarly, the idea that there is in the universe no fixed point or frame of reference, with the consequence that there is no absolute motion, contradicts intuition. Other examples from physics could be cited.

There has also been a downfall of intuition in mathematics. Godel's undecidability theorem, which says that for most logical systems statements can be made which cannot be proven or disproven within the system, is but one example of a mathematical conclusion which is contrary to what was expected intuitively.

The idea of no beginning is an idea which man intuitively rebels against. Every process in nature which he observes suggests beginning, growth, and end. Yet, since intuition is not reliable in *many* important instances, the use of intuition in *any* scientific problem is therefore suspect. The use of intuition in science and analytical thought in general has been under widespread attack. It is precisely this development which has made it possible and scientifically respectable for the natural man to suggest that matter may be eternal.

The natural man's inability to prove the existence of an adequate cause-and-effect law is ironically something which he attempts to turn to his advantage in the question concerning the existence of a beginning. If his intuition breaks down one place, he reasons, it can also break down when it tells him that there must have been a beginning. A fundamental goal of the natural man is to comprehend, in principle, the entire universe. Pursuit of this goal has led him into difficulties, one of which is the problem of cause and effect, a problem which troubles the foundation of the science of the natural man. Thus, the natural man uses the breakdown of intuition to solve the chief of all his problems, the problem of a beginning.

It might be objected that it is intuition, and not reason, which has collapsed, and that, after all, it is *reason* upon which the natural man depends. Is it possible that one can say, for example, that men expected *intuitively* that there can be absolute motion, while Einstein showed by brilliant *reasoning* that this intuition was wrong? Similar questions could be asked about the other examples showing the downfall of intuition. However, Einstein or whoever breaks down previous intuition utilizes in his refutation some hypotheses which are unproved. They who reason thus admit freely that they use unproven hypotheses. Since intuitions have been shown to be unreliable, it has also been shown that we cannot depend ultimately upon our basic hypotheses, which themselves are intuitions.

Consider how some of these ideas have been developed in this book. Men encounter a basic difficulty when they attempt to decide which statements in the Bible are true, and this basic difficulty arises because they have *no yardstick*, no criterion which can be put above the Bible. When men accept no absolutes from God, they have no means – again, no yardstick – to enable them *ultimately* to decide certain physical problems, such as whether or not the universe is finite. Similarly, there has been a sufficiently successful attack on intuition during the last century to render suspect any attempt to differentiate between acceptable and unacceptable hypotheses. Man can produce no yardstick which enables him to differentiate between acceptable and unacceptable hypotheses.

Thus, while there has been fantastic progress in analytical thought, the natural man has been unsuccessful in his attempts to produce his own yardsticks. He has convinced himself of the truth of Propositions 2-5, and the next logical step has been the acceptance of Proposition 1 which postulates the eternity of matter. Many scientists have concluded that matter is indeed eternal, and with the situation as it is, in time many more scientists will surely come to the same conclusion.

There is a cloud over the reasoning of the natural man. He desperately needs the concept of no beginning. He realizes that if there were one instance of creation, there could have been others, and his whole system would collapse. He sees that the only way he can accept the concept of no beginning is that he depend upon the downfall of intuition. For him, this is an awful choice. To avoid the ultimate in his problems, the problem of creation, he must lean on the downfall of intuition, a downfall which is actually the unreliability of his own reasoning. *But his purpose in attempting to avoid the idea of creation or of a beginning is to satisfy his goal of complete comprehension.* Therefore, there is a basic contradiction in the thinking of the natural man.

Suppose the natural man can find a personally satisfying way out of his dilemma. Suppose that he can show to his satisfaction (even though the idea would be contrary to what the Bible teaches) that matter could be eternal *without* putting the whole problem in the category of it-must-be-true-although-it-cannot-be-proved. The question would then not be in the mysterious region of unproven ideas which are contrary to our intuition. The natural man cannot, however, debate this matter on the basis of what he *might* develop. His position is such that he cannot depend upon future developments. For the present and foreseeable future, he has reasoned himself into a corner he did not intend to enter.

The Christian Reply

The natural man has therefore left the foundations of his science in a shambles. He has tried to change the rules of the game, and yet he still cannot achieve his goal. In attempting to obtain an adequate law of cause and effect, he has tried to change the rules by relying upon the statistical prediction for individual particles. He knows, however, that he has always wanted to be unrestricted in his ability to make predictions within the physical universe. With the problem of the concept of a beginning, he has attempted to use man's demonstrated weakness, a tendency to choose some wrong first principles, to achieve a victory for man's reasoning ability, the ability which by definition should include the ability to choose correct first principles.

The Christian must attempt to make order out of the shambles by attacking the very problems which give the natural man difficulty. With respect to the problem of cause and effect, the Christian answers that not all causes are contained within the universe. For the Christian, the problem of cause and effect is but one indication that there is a cause outside the universe. With respect to the concept of a beginning, the Christian maintains that (1) there was a beginning, since the Bible teaches it; (2) a beginning implies a creation; (3) a creation implies a creator; and (4) the existence of a creator implies there is a cause outside the universe. Thus the Christian answer to both of these fundamental problems of the natural man is the same. Ultimate causes lie outside the universe, and the universe is not self-contained. *Therefore, man cannot fully comprehend the universe even in principle.*

The Christian does not maintain there was a beginning merely because he sees birth, growth, and death in nature. He accepts the concept of a beginning because the Bible tells him there was a beginning.

The Christian knows that there can be a breakdown of intuition, and the Christian fundamentally knew this long before modern science demonstrated many of these breakdowns. Concerning a beginning the Bible validates man's intuition. The passage quoted above (Ps. 102:25-27) tells of the laying of the foundation of the earth, of the waxing old, and of the perishing – in other words, it tells of birth, growth, and death.

What has been said here has been an argument for the existence of a creator from the idea of a first cause. Even so, it is improper to maintain that the idea of a first cause is *correctly* deduced by the mind of man. Intuition can break down, and the mind of man can make an error. In fact, it seems that the traditional cause-and-effect first cause argument for the existence of God could not be used today, since man cannot prove the existence of an adequate law of cause and effect.

Thus, the Christian's answer is always found in God's inscripturated revelation, the Bible. Even without the Bible, the inadequacy of the mind of man to comprehend the universe can be shown, and the existence of something beyond the universe is suggested. But, any conclusion not using the Bible must in its very nature be tentative. The Christian knows the God who is beyond the universe because he possesses absolute truth in the Bible. Through the Bible men know God as the one who is outside of the universe he created, a universe which therefore had a beginning. Thus, when man speaks of time he always means "time in our universe," and it follows that God exists outside of time as well as outside the universe. It is not the universe which has existed from eternity, but it is God who is uncreated and eternal. The universe does not contain its ultimate causes, but God, outside the universe, is the source of ultimate cause. Referring back to the question of the size of the universe, it is not the universe, but the God who created space, who is infinite. With respect to the other basic problem of the natural man discussed earlier, that of rejecting or accepting evidence, it is only the believer in God and God's revelation who can meaningfully discuss criteria for accepting and rejecting evidence.

In other words, the Christian of himself cannot *comprehend* any better than the natural man. The natural man cannot comprehend ultimate causes and the problems of finitude; but the Christian, while he knows that God is infinite with respect to the power, space, and time which men know, cannot go beyond that. The natural man cannot solve the problem of the origin of the universe; but the Christian cannot understand how God can be uncreated.

In all this the Christian is content, however. He knows what he cannot know. He knows that the universe depends upon God. On the other hand, the natural man is today showing more clearly that his basic assumption contains internal contradictions, and cannot be true in this universe. Because the Christian has the correct answer, he is content; because the natural man has the wrong answer, an answer that does not correspond with reality, he is not content.

REFERENCES

1. Calvin, J., *Institutes of the Christian Religion*, Vol. I, Eerdmans, Grand Rapids, Mich., 1949; p. 67.

9 The Bible on the Age of the Universe

The second division of the discussion of evolution is Proposition 2, "The earth and at least some stars are billions of years old." This proposition is an essential part of the theory of biological evolution, since billions of years are needed for the biological changes postulated. Biological evolution (Propositions 3-5) is discussed in later chapters.

Is it possible that the first creative act of God occurred billions of years ago? Some Christians believe that all creative acts took place within six 24-hour days thousands of years ago, whereas others believe that the universe is billions of years old, and that at least some of the days of Genesis 1 were long periods. According to this belief, "day" was used by Moses to denote an indefinite period of time in the sense that one might say, "In Lincoln's day the country was in turmoil."

Sometimes those who debate this issue are not agreed upon just what the issue is. For this reason, two preliminary explanations must be made. First, the concept of a day as a long period does not imply that there was a long period of light, followed by a long period of darkness. "Day" in this interpretation does not mean "long day." "Day" refers to a long, somewhat indefinite period of time in the sense that "Lincoln's day" is indefinite. Similarly, "evening" and "morning" in Genesis 1 are the corresponding parts of this long-period "day." "Evening" and "morning" are used as similar terms are used in "on the eve of the French Revolution," the "twilight of his life," and "at the dawn of history."

Second, in discussing a short or a long period for each day of Genesis 1, the ability of God to create in a short period is not in question. Christians maintain that God could create everything in one instant. Yet, no one maintains that he did create everything in a single instant. If speed is the essence of the miraculous, then God's creation was not as miraculous as it could have been. It seems that the proper position to take is that speed is not the essence of the miraculous, and that creation does not become more miraculous by being completed in a short time.

"Day," a Textual Problem

It is assumed here that the six creation days of Genesis 1 were chronological days. (A different assumption, probably contrary to the text, has been made that the six days refer to a framework of creation acts and not to chronological acts; space prohibits discussion of this idea.) If the days were chronological, then it is their length which is at the heart of the question concerning the age of the universe.

It is often stated that the Bible is to be taken literally, and that therefore "day" in Genesis 1 refers to either a 24-hour day or to something very close to that. The Bible is indeed to be taken literally, but the literal meaning of a passage is sometimes difficult to obtain. Understanding "day" in Genesis 1 is not easy, even though the word is to be taken literally. There are two reasons why the meaning of "day" in Genesis 1 is not immediately obvious from the text.

A. *The Biblical Description of the Creation Days.* From the Biblical description of the creation days, one cannot deduce how long these days were. Creation was miraculous, but it cannot be stated how rapidly the events occurred. Some statements of Genesis 1, such as, "Let there be light," seem to refer to instantaneous acts; others, such as "Let the earth put forth grass," seem to be statements which could refer to either an instantaneous act or a slow process. Furthermore, even if it is assumed that all creative acts were instantaneous, it would not be known how much time elapsed between the various creative acts.

B. *The Use of* Yom. The other reason there is difficulty in ascertaining the meaning of "day" in Genesis 1 is that the Hebrew word used here, *yom,* which appears hundreds of times in the Bible, can be used to denote periods of different length. These uses range from the daylight period of the day up to long periods.

In a passage such as the following, *yom* means solar day, a certain calendar day:

> And the ark rested in the seventh month, on the seventeenth day of the month, upon the mountains of Ararat. (Gen. 8:4)

Yom is used similarly in many other passages, such as,

> Then shalt thou cause the trumpet of the jubilee to sound on the tenth day of the seventh month, in the day of atonement shall ye make the trumpet sound throughout all your land. (Lev. 25:9)

In some passages *yom* refers not to a solar day, in the sense that a definite day of the calendar is referred to, but rather to a time or a period of time. Job, speaking of man's life, provides an example:

> Turn from him, that he may rest, till he shall accomplish, as an hireling, his day. (Job 14:6)

Jeremiah, in prophesying doom for the Chaldeans, equates *yom* to a period, a time of visitation:

> Slay all her bullocks; let them go down to the slaughter: woe unto them! for their day is come, the time of their visitation. (Jer. 50:27)

Yom can also denote a timeless period:

> Thou art my son, this day have I begotten thee. (Ps. 2:7)

Some of the other passages in which *yom* is used to denote a time other than one solar day: Gen. 2:4; Deut. 1:10, 31:18, and 34:6; Job 18:20; Ps. 37:13 and 137:7; Prov. 4:18; Isa. 61:2, 63:4, and 65:2; Jer. 30:7; Lam. 2:22; Ezek. 7:10, 21:25, and 39:8; Hos. 1:11; Joel 1:15, 2:1-2, 2:11, 2:31, and

3:18; Micah 3:6; Zeph. 1:14; Zech. 2:11, 4:10, and 14:7; Mal. 4:1. (Some of these passages are discussed later.)

Long-Period Creation Days

Evidence to indicate that *yom* in Genesis 1 refers to 24-hour days is therefore lacking. (Some objections to this conclusion are discussed later in this chapter.) In addition, Biblical evidence suggests that the creation days were indefinite periods of time. This evidence is divided into two categories.

A. *God's Activities and* Yom. It is of interest to consider the nature of God's activities, as recorded in the Bible, which are associated with *yom.* When God deals with nations, the wicked, or his people; when he describes a turning point in history, such as Christ's work on earth or the end of time; or when he speaks of the whole act of creation; *then yom* is used to refer to a period longer than a 24-hour day. It can denote a period of many years. Some examples of these uses of *yom*, uses associated with God's work, are now given.

God deals with his people:

> And it shall come to pass in that day, that the mountains shall drop down new wine, and the hills shall flow with milk, and all the rivers of Judah shall flow with waters, and a fountain shall come forth of the house of the Lord, and shall water the valley of Shittim. (Joel 3:18)
>
> And many nations shall be joined to the Lord in that day, and shall be my people: and I will dwell in the midst of thee, and thou shalt know that the Lord of hosts hath sent me unto thee. (Zech. 2:11)

God deals with the wicked:

> The wicked plotteth against the just, and gnasheth upon him with his teeth. The Lord shall laugh at him: for he seeth that his day is coming. (Ps. 37:12-13)
>
> And thou, profane wicked prince of Israel, whose day is come, when iniquity shall have an end. (Ez. 21:25)
>
> Thou hast called as in a solemn day my terrors round about, so that in the day of the Lord's anger none escaped nor remained: those that I have swaddled and brought up hath mine enemy consumed. (Lam. 2:22)

God causes a turning point in history:

> Alas for the day! for the day of the Lord is at hand, and as a destruction from the almighty shall it come. (Joel 1:15)
>
> Behold, the day of the Lord cometh. . . . Then shall the Lord go forth. . . . And his feet shall stand in that day upon the Mount of Olives. . . . And it shall come to pass in that day, that the light shall not be clear, nor dark: [i.e., it will be light over the whole earth at the same time] but it shall be one day which shall be known to the Lord, not day, nor night: but it shall come to pass, that at evening time it shall be light. And it shall be in that day, that living waters shall go out from Jerusalem; half of them toward the former sea, and half of them toward the hinder sea: in summer and in

> winter shall it be. And the Lord shall be king over all the earth: in that day shall there be one Lord, and his name one. (Zech. 14:1, 3, 4, 6-9)

All these grand events which Zechariah depicts are associated with "one day *which shall be known to the Lord*" (italics added). The Lord's day is not necessarily a solar day. Even though the day described is said to be a day with which light and darkness cannot be associated in the ordinary way, it is nevertheless a day which has an "evening time." It is therefore possible to speak of an evening (and, presumably, a morning) of a long period — *yom.* It is difficult to understand how a Bible-believer can insist, after having studied this passage, that God's grand activities of creation week *must* have occurred in six solar days.

God created all things:

> These are the generations of the heavens and of the earth when they were created, in the day that the Lord God made the earth and the heavens. (Gen. 2:4)

There is no hint in the Bible that God *needs* periods of time longer than 24 hours to bless his people, vanquish the wicked, accomplish the work of Christ, or culminate history. These are grand events which men could expect to be accomplished over a period of time, a period which could be as long as many years. Furthermore, *yom* in "This day have I begotten thee" (Ps. 2:7) is timeless. All these grand events, activities of God, are said to take place in a day, where the same Hebrew word is used as is used for each of the six creation days on which other grand works of God occurred.

Naturally, not all the grand events recorded in the Bible took place over periods of time longer than 24 hours. The flood ended on a particular solar day; Israel was given key military victories on certain solar days. On other solar days the Israelites were punished with key defeats, such as the battle in which King Josiah was killed. The Word became flesh on a certain solar day, and Christ rose from the dead on another.

The important fact is that many events which occurred on a "day" took long periods of time. If Bible-believers could forget completely what tradition says about the creation days, as they ultimately must, they would almost certainly conclude that the Bible itself gives no indication that all six days were 24-hour days, and that at least some of them were long periods.

B. *The Sabbath.* With this idea of what "day" *(yom)* often means when it is associated with God's activity, the creation week must now be re-examined. At the end of the creation "week" God rested:

> And on the seventh day God ended his work which he had made; and he rested on the seventh day from all his work which he had made. And God blessed the seventh day, and sanctified it: because that in it he had rested

> from all his work which God created and made [i.e., created to make]. (Gen. 2:2-3)

Here also *yom* is translated "day." Five comments concerning the sabbath are made.

(1) Since the seventh day is discussed in Genesis in parallel with the six days, there seems to be a parallelism with respect to the length of these days. If the seventh day was a solar day, it would be straining exegesis to hold that the six days were not solar days. If the seventh day was longer than a solar day, it would seem likely that the six days were also longer than solar days. If we can learn something about the length of the seventh day, we can almost certainly know something about the length of the six creation days.

(2) God's seventh day of rest after six days of work is the Biblical warrant for man's working six solar days and resting the seventh solar day. This commandment is implied in the passage just cited ("And God blessed the seventh day and sanctified it") and it is stated explicitly in the Fourth Commandment:

> Remember the sabbath day, to keep it holy. Six days shalt thou labor, and do all thy work: but the seventh day is the sabbath of the Lord thy God: in it thou shalt not do any work. . . . For in six days the Lord made heaven and earth, the sea, and all that in them is, and rested the seventh day: wherefore the Lord blessed the sabbath day, and hallowed it. (Ex. 20:8-10, 11)

(There is also a description of man's sabbath and God's sabbath in Exodus 31:13-17.)

In view of this passage, is it permissible to conclude that God's sabbath and creation days were anything other than solar-length days? It might at first seem that if God's creation week consisted of days of indefinite length, man would thereby have permission to work and rest in the same pattern of six periods followed by one period with each period of indefinite length. Since man does not have such permission (indicated in the Bible in many places, and even known to the Israelites before the law was given (Ex. 16:5, 22-30), it would then follow that God's creation week consisted of solar-length days.

But there is one question which must be answered before such an argument can be adopted. Did the Israelites know, and can we know, that God's sabbath was a period longer than a solar day? If the Israelites knew God's sabbath to be such a period, and if we can know the same, then the Fourth Commandment must be understood in *this* context. The commandment would then tell men that even as God worked on six of God's days and rested on the seventh of God's days, so should man work on six of man's days and rest on the seventh of man's days.

The Israelites could have known God's day of rest to be a long period if they considered God's own definition of his rest, his sabbath. They were told in Genesis 2:2-3 that his rest was a *cessation of creation.* What happened on the six creation days was in some way different from what happened at any later time. Many miracles God performed later were kinds of creation; yet, in no way does the Bible indicate that there was ever again a time which was like the time of the creation week. God created then, and he now upholds his creation by his providence. After the creation week, there was no further creation, but the upholding, the providential guidance which had already begun, was continued and continues until now.

This change – from a time of creation and providential guidance to a time restricted to providential guidance – occurred, according to Genesis 2:2-3, when God's sabbath began. For this reason, it is concluded that God's sabbath was not a solar day, but that it is a long period. Since Genesis 1 and 2 and Exodus 20 teach a parallelism between the six days of creation and the seventh day, it is concluded that the six creation days were also periods which had no special relationship to the solar day, although the possibility that some of them were solar days is not eliminated.

(3) The Israelites were to keep other sabbaths besides the weekly sabbath. An examination of these other sabbaths leads almost certainly to the conclusion that God's sabbath cannot be limited to a solar day:

> . . . When ye come into the land which I give you, then shall the land keep a sabbath unto the Lord. Six years thou shalt sow thy field, and six years thou shalt prune thy vineyard, and gather in the fruit thereof; but in the seventh year shall be a sabbath of rest unto the land, a sabbath for the Lord: thou shalt neither sow thy field, nor prune thy vineyard. . . . It is a year of rest unto the land. And the sabbath of the land shall be meat for you. . . And thou shalt number seven sabbaths of years unto thee, seven times seven years; and the space of the seven sabbaths of years shall be unto thee forty and nine years. . . . And ye shall hallow the fiftieth year. . . . A jubilee shall that fiftieth year be unto you: ye shall not sow, neither reap that which groweth of itself in it. . . . (Lev. 25:2-4, 5, 6, 8, 10, 11)

Thus, the six-one pattern was used when the unit was not the solar day, but the year. The seven-one pattern was also used, with the unit being seven years, while the fiftieth year was the jubilee year. It is possible that the seven-one pattern was used because out of the forty-nine years, only forty-two years, i.e., six times seven years, were work years. Thus if the one-year sabbaths are not counted, there is the six-one pattern here also.

At least one of these Levitical work-rest patterns, given in terms of years, was patterned after God's work-rest creation week pattern. The Israelites would have been puzzled if they had thought the Fourth Commandment taught that God's creation and rest days were just as long as man's work and rest days. Where, then, would the *yearly* six-one pattern fit in? The dilemma is removed if the Israelites understood that they were not to make a time analogy between the days of God's creation week and either the Israelite work week or the six-year period the land was to be worked.

(4) There have been objections to the sabbath argument used here. It has been suggested that God's sabbath was a solar day of rest which was followed by more work. In support, Christ has been quoted:

> But Jesus answered them, My Father worketh hitherto, and I work. (John 5:17)

It is inconceivable that God did not work in this sense *immediately* after the end of the sixth creation day. God's providence could not be suspended one instant, much less an entire solar day. God's providential work since the moment of the first creative act has never ceased. The description of the sabbath in Genesis 2:2-3 has therefore often been misunderstood. The passage does not state that God rested from all work. This passage says three times that God's sabbath is a cessation of *creative* work: (1) ". . . on the seventh day God ended his work which he had made"; (2) ". . . he rested on the seventh day from all his work which he had made"; (3) ". . . in [the seventh day] he . . . rested from all his work which God created and made."

Thus, if one holds to a solar rest day for God there is a consequence that is not often considered. His rest would therefore have ended at the end of this solar day, and one would be forced to conclude that on the eighth day God *resumed* creating. Surely, such a conclusion cannot be reasonably held. God's creative work, *which he pronounced good after six creation days,* was then finished. God's sabbath has continued from then until now.

Is it possible that these passages associating God's sabbath of rest with the seventh day merely mean that God's rest *began* on the seventh solar day, and that his rest *continued* on the eighth, ninth, etc., days? In these passages God's resting from creation defines his sabbath. His sabbath *is* his seventh day. These three concepts cannot be dissociated from each other: God's rest, God's sabbath, and God's seventh day. For God, there are apparently no eighth, ninth, etc., days.

(5) There is a New Testament discussion of the institution of the sabbath.

> For we which have believed do enter into rest, as he said, As I have sworn in my wrath [referring to Num. 14:23 and Ps. 95:11, speaking of the disobedient Israelites who were in the desert for forty years], if they shall enter into my rest: although the works were finished from the foundation of the world. For he spake in a certain place of the seventh day on this wise, And God did rest the seventh day from all his works. And in this place again, If they shall enter into my rest. . . . There remaineth therefore a rest to the people of God. For he that is entered into his rest, he also hath ceased from his own works, as God did from his. Let us labor therefore to enter into that rest. . . . (Heb. 4:3-5, 9-10)

This passage apparently teaches that God's works were finished at "the foundation of the world," that his seventh day of rest is the rest which follows those works, and that his rest still continues.

The "day" of "seventh day" in this passage is given by the Greek *hemera.* It cannot be deduced from the meaning of *hemera* whether a solar day or a long period is meant, since this word can have either meaning in the New Testament. Following are some of the many instances in which *hemera* cannot mean "solar day":

> The sun shall be turned into darkness, and the moon into blood, before that great and notable day of the Lord come. (Acts 2:20)
>
> But after thy hardness and impenitent heart treasurest up unto thyself wrath against the day of wrath and revelation of the righteous judgment of God. (Rom. 2:5)
>
> As also ye have acknowledged us in part, that we are your rejoicing, even as ye also are ours, in the day of the Lord Jesus. (II Cor. 1:14)
>
> Not forsaking the assembling of ourselves together, as the manner of some is; but exhorting one another: and so much the more, as ye see the day approaching. (Heb. 10:25)

It is not denied here that history will culminate on a single day. But the New Testament teaches that many events will lead up to that last day, and it seems certain that "day" in these passages refers not to the last solar day, but to the "end time," a "day" which is a period.

In summary, the Bible does not seem to indicate anywhere that God's sabbath was a solar day. On the contrary, it is very likely that the Bible teaches that God's sabbath is a long period, and that at least some of the six creation days are likewise to be considered long periods.

Questions Concerning the Interpretation of "Day"

A. *The Israelites' Idea of "Day."* It was suggested above that the Israelites had good reason for not associating God's creation days with solar days. Suppose, for the sake of the discussion that this suggestion is incorrect, and that the Israelites did indeed believe the creation days to be solar days.

However, the Bible is not to be interpreted assuming the Israelites' opinions are normative. There were surely many incorrect ideas about the universe among all the groups who first heard God's Word – the Israelites, the churches to whom the epistles were written, the dispersed Christians, etc. Moses and the authors were guided to write correctly, and it may be assumed that these authors wrote so that the first readers could have understood correctly. Errors on the part of the first readers of the Bible should not affect present understanding of the Bible. Perhaps this can be understood more easily by considering the Old Testament prophecies of Christ's life on earth. Those prophecies were certainly not thoroughly understood by the first readers.

What must be avoided is reading into the Biblical text a modern definition of an ancient term. When Moses wrote of "kinds" in Genesis 1, he did not necessarily mean "species," a term given its modern meaning only a few hundred years ago. When Peter wrote of "elements" (II Peter 3:10), he did not necessarily refer to what is now meant by that word.

God inspired Moses to write not only to the Israelites, but also to all of God's people in succeeding ages. If the cultural ideas of the readers are to be taken as normative, it can well be asked, "*Which* readers?" Were we to depend wholly on *our* science, thus using the cultural ideas of the present day, we would take the days to be long periods. Neither present-day science nor the ideas that the Israelites had may be normative in determining the meaning which God intended as he inspired Moses to write "day."

B. *Numbered Days.* Wherever numbered days are mentioned in other parts of the Bible, they are solar days. Therefore, it has been argued that the creation days, which also are numbered, were also solar days.

Wherever "day" refers to a period of time obviously longer than a solar day in other parts of the Bible, there is *no need* to number the day. The "day of the Lord," the "day" on which the Father begat the Son, the Lord's "day of vengeance," etc., are all days which do not require numbering. On the other hand, the solar days referred to sometimes required numbering; for example, on the eighth day the child was to be circumcised (Lev. 15:24). If God created in six chronological days of indefinite duration, it is very likely he would number the days chronologically. This would then be the one place in the Bible which refers to chronological long period days, and, logically, the one place in which long period days are numbered.

C. *Events of the Creation Days.* Another objection to accepting any of the creation days as long periods is based on the idea that the creation acts themselves took place instantaneously. It is maintained that some of the words in the Bible used to describe creation days suggest suddenness. An example of a passage quoted to prove this point is part of Psalm 33:

> By the word of the Lord were the heavens made; and all the host of them by the breath of his mouth. . . . For he spake, and it was done; he commanded, and it stood fast. (Ps. 33:6, 9)

There is no doubt that *each creation event* was instantaneous. One moment a certain thing existed; the previous moment, it did not exist. On some creation days there may have been many such individual creation events, with unknown time lapses between these events. Passages such as the one cited from Psalms probably refer both to the initial, instantaneous event and also to subsequent developments.

Death Before the Fall

One of the most frequent objections to equating the creation days with long periods is that long periods of time prior to Adam are not possible because, it is said, there was no death in nature prior to Adam's fall. Adam's fall is claimed to have brought death not only to man but also to nature. With no death in nature, it is inconceivable that plants and animals could have lived and been fed for the long periods which may have continued for millions of years. It follows that plants and animals were created a very short time before Adam fell and brought a curse upon creation.

However, the death the Bible ascribes to Adam's sin is the death of men. Paul says,

> Wherefore as by one man sin entered into the world, and death by sin; and so death passed upon all men, for that all have sinned. (Rom. 5:12)

Paul neither affirms nor denies that Adam's sin brought all death into the world.

The position adopted here is that the Bible nowhere teaches that Adam's sin brought death to nature, but that man's sin did in some way affect nature. How was nature affected? When Adam sinned, he declared that he could function quite well without God. As punishment, God no longer completely protected him from his natural environment. He could now fall and break a leg. Certain bacteria, created during the six creation days, could now harm him. Adam was told that because he had sinned,

> . . . Cursed is the ground for thy sake; in sorrow shalt thou eat of it all the days of thy life; thorns also and thistles shall it bring forth to thee; and thou shalt eat the herb of the field; in the sweat of thy face shalt thou eat bread, till thou return unto the ground. . . . (Gen. 3:17-19)

The conditions which had existed before the fall were thus allowed to affect man because of his sin. The thorns and thistles, which had been in existence since the creation days, God now allowed to annoy and hinder man.

A. *Biblical Indications of Death Before the Fall.* There are at least three indications that there was death, other than human death, prior to Adam's fall.

(1) Adam and Eve were permitted to eat:

> And the woman said unto the serpent, We may eat of the fruit of the trees of the garden. (Gen. 3:2)

The eating which was allowed caused the death of fruit cells and bacteria in the digestive tract. There is no reason to suppose that Adam's eating differed from our own. Since the Bible speaks to those who live after Adam's fall, it should be assumed that the Bible means "eating" in the sense that post-fall man understands this term. Why should deep mystery be ascribed to eating?

(2) Consider Genesis 1:28-30:

> And God blessed them, and God said unto them, Be fruitful, and multiply, and replenish the earth, and subdue it: and have dominion over the fish of the sea, and over the fowl of the air, and over every living thing that moveth upon the earth. And God said, Behold, I have given you every herb bearing seed, which is upon the face of all the earth, and every tree, in the which is the fruit of a tree yielding seed; to you it shall be for meat. And to every beast of the earth, and to every fowl of the air, and to every thing that creepeth upon the earth, wherein there is life, I have given every green herb for meat: and it was so. (Gen. 1:28-30)

This passage might at first be thought to prohibit eating meat for both men and animals. Although the statement is positive concerning plants it does not necessarily follow that this passage prohibits eating meat.

Eating meat would of course cause death. There are many animals that were created with claws and large, sharp teeth, giving them the ability to kill other animals and tear flesh for eating. One would hardly expect claws and fangs to be used on vegetation alone. It is a peculiar view of death which distinguishes between the death of plants and the death of animals.

Furthermore there seems no evidence that before the fall man was prohibited from killing animals for purposes other than food. Man is given dominion over all the animals. Thus, man may cut open a frog to determine its structure – and such researches enable him to improve man's health and comfort – or man may eat meat.

If having dominion meant something different to Adam before the fall than it does to us, then one must conclude that the entire pre-fall account is ambiguous. One could then counter the interpretation of any part of the pre-fall account with, "Sin changed all this." Such an attitude, an attitude of ascribing mystery merely in order to fortify a previously-held idea, is close to the attitude of those who debunk much of the Bible to suit their own ends. If the Christian considers the first two chapters of Genesis to be inherently beyond our comprehension, then the door is left open for the wildest ideas about creation.

(3) Another Biblical indication that there was death other than human death before Adam's fall, is that Adam must have known before the fall what death is, since Eve spoke to the serpent concerning the forbidden fruit prior to her sin:

> . . . God hath said, Ye shall not eat of it, neither shall ye touch it, lest ye die. (Gen. 3:3)

If an inerrant Bible is accepted, it follows that these words were spoken, and that Adam, Eve, and the serpent all knew what "lest ye die" means. Adam and Eve knew about death because they saw it in their environment. God protected them from what this environment could to to them as long as they trusted completely in him. When God said, "lest ye die," he was threatening that he would withdraw some of this protection if they withdrew their trust.

In summary, processes of nature imply death; and there seems to be no Biblical evidence to indicate nature was different before the fall.

A Groaning Creation

An objection to the conclusion just made is the claim that Adam's sin brought a curse, not only upon Adam and his descendants, but upon all creation, and that this curse included death. The following passage has been cited:

> For the creature was made subject to vanity, not willingly, but by reason of him who hath subjected the same in hope, because the creature itself shall also be delivered from the bondage of corruption into the glorious liberty of the children of God. For we know that the whole creation groaneth and travaileth in pain until now. And not only they, but ourselves also, which have the firstfruits of the Spirit, even we ourselves groan within ourselves, waiting for the adoption, to wit, the redemption of our body. (Rom. 8:20-23)

It has been held that groaning and travailing in pain are a consequence of the corruption mentioned, that groaning and travailing imply death for all animate life, and that corruption is punishment for Adam's sin, his "vanity," v. 20. In this approach, this passage is associated with the curse given to Adam (Gen. 3:17-18, quoted above), in which he was told that the ground was cursed for his sake and he would be plagued by thorns and thistles.

Much of this analysis seems to be correct. All animate life is apparently involved in the pain Paul refers to. The corruption Paul speaks of has some relation to the curse spoken to Adam. But is it possible to deduce from this passage that all death in nature is a consequence of Adam's sin? The pre-fall animate creation of which Paul speaks is a creation in which the life cycle existed. The creation which was affected by Adam's sin was a creation in which Adam ate, thistles existed, animals had claws and sharp teeth, and Adam knew what "lest ye die" means.

Creation could have been cursed because of man's sin in ways other than the introduction of death. There is no need to imply that we have full understanding of this Romans passage, but it is possible that other parts of the Bible can aid us in an attempt to understand Paul's meaning. Consider these two passages:

> He turneth rivers into a wilderness, and the watersprings into dry ground; a fruitful land into barrenness, for the wickedness of them that dwell therein. (Ps. 107:33-34)
>
> The earth also is defiled under the inhabitants thereof; because they have transgressed the laws, changed the ordinance, broken the everlasting covenant. Therefore hath the curse devoured the earth, and they that dwell therein are desolate: therefore the inhabitants of the earth are burned, and few men left.

> The new wine mourneth, the vine languisheth, all the merry hearted do sigh. (Isa. 24:5-7)

Adam's sin polluted the whole race, making all men sinful. Frequently in man's history his sin brought a curse upon the land. Probably the devastation these passages refer to is at least part of the groaning-and-travailing idea of which Paul writes.

An indication that the curse on creation was not instantaneous, such as it would have been had the curse consisted only of the introduction of death, is the slow decrease in the length of man's life, as indicated by the Biblical record of the ages at which men died. This slow decrease, which continued for a long time after Adam lived, must nevertheless have been a part of the curse on creation.

Thus, there are several indications that the general curse on creation consisted of at least some elements other than the introduction of death. It is therefore possible that Paul referred only to these other elements when he referred to groaning, travailing, and corruption. There seems to be no reason to insist that Paul taught that man's sin brought all death into creation.

Summary

What has been attempted in this chapter is to show that many apparently pious ideas concerning what is discussed in the first chapters of Genesis do not rest on what the Bible says. Thus, the Bible admittedly uses "day" for either short or long periods. In spite of this, it is often assumed without proof that the idea of a short period can be imposed on Genesis 1. Paul and Moses both state that man's death came because of Adam's sin. This idea is often extended, again without proof, to include all death. The Genesis account states that man was told he could eat of the fruit of plants. This is often interpreted to mean that man was not allowed to eat meat, even though the Bible makes no such statement. Some Biblical miracles occurred suddenly. It is often unjustifiably deduced from this idea that the miracle of creation could involve only suddenness and no process. Some Genesis 1 and 2 passages seem to refute these pious, but probably incorrect, ideas. The answer which is often given – again, at first sight a pious answer – is that sin, which appeared in Genesis 3, changed everything, so that Genesis 1 and 2 cannot be used in this way. But the Bible nowhere teaches that we may treat Genesis 1 and 2 so casually.

While only some of the interesting questions concerning the days of creation have been discussed, an attempt has been made to include several pasages which can be helpful in considering the length of the creation days. It is concluded that the Bible suggests that the creation days were not necessarily solar days, and that it is possible that at least some of them were long periods. Only at this point may one turn to science with an open mind and ask, "How long were the creation days?" or, to ask the more common question, "How old are the stars and the earth?"

10 Science on the Age of the Universe

If the conclusions of the last two chapters are correct, the Bible indicates that at least some of the creation days may have been long periods. Questions about the *age* of the earth and the stars, not about creation days, are asked of science. Naturally, a great age implies that there were some long creation days, while a very young age could be consistent with the solar creation day concept.

In science there are levels of certainty, and it is unwise to doubt the essential correctness of science in those large areas of science where scientists agree. The "level of certainty" concept is usually ignored when demonstrated scientific error in one matter is cited in an effort to debunk a large part of science. The caution suggested concerning this "level of certainty" is particularly important in estimating the age of creation. Some methods of obtaining the age of the earth, the other planets, and the stars are understood by scientists to be crude methods, and only of secondary value. It is not fair to use the crudity of these methods in an effort to reject the conclusions based on better methods. The best methods, the one derived from astronomical conclusions and the one using uranium, thorium, and potassium radioactivity, are the methods to scrutinize.

Astronomy and Age

A. *The Age Consequences of Astronomical Conclusions.* Stars can be observed only by virtue of the light or other radiation which they emit. If a star is one million light years away, we observe light which left that star one million years ago. (Naturally, it is possible that God created the light on the way from the stars at some time less than one million years ago. The apparent age would then be greater than the real age. The important matter of apparent age arises also in connection with other methods of determining age, and consequently the apparent age question is discussed later along with other objections to the methods of determining age.)

The idea of stars or galaxies millions or billions of light years away causes one to marvel at the great distances there are in the universe. Since light travels 186,000 miles per second, these distances are indeed amazing. However, the concepts of great distance and great age cannot be separated. Acceptance of these great distances implies acceptance of the idea that the first creation acts took place billions of years ago. Suppose for the sake of the discussion it is assumed (1) that the first acts of creation took place ten thousand years ago, and (2) that this approximate date for creation can be deduced from the Bible. Such a position is a probable consequence of taking the creation days to be successive solar days.

If the Christian believes that God tells him in the Bible that creation occurred approximately ten thousand years ago, the Christian then has *no proof* that anything exists farther than ten thousand light years away. If God has revealed the approximate date of creation, it makes no difference that most of the universe seems to be much farther away than the ten thousand-light year limit. It might be objected, "All we need do is assume that God created the light on its way. God is faithful, and if certain starlight indicates a star is a million light years away, it means there is such a star, even though we have not yet seen light which originated at the star." If God created but ten thousand years ago, and if he says this in the Bible, man has no right to assume that there is a million-light-year-distant star "behind" the light now seen, light which this view assumes was created relatively close to the earth only ten thousand years ago. If God tells man he created this light *en route*, the knowledge of such a creation is at least as reliable as knowledge derived from the astronomer's calculations of apparent distance. The Christian then ought to insist that while the term "apparent distance" is acceptable, such a distance has no relation to an actual distance. Not even the existence of the star would have been proved for the Christian.

Consideration of a more familiar means of determining age also demonstrates the problem. If God revealed to Adam at a certain time that the first creative act had occurred just one year earlier, and if a few days later Adam cut down a tree exhibiting one hundred rings, should he have dared to assume creation was at least one hundred years old? If Adam had believed God's word, he would have understood that the tree had an apparent age of one hundred years, and that this apparent age bore no relation to its actual age. It would have been proper to assume that the tree had been created with rings. If one of the created rings of Adam's tree was thin, would that indicate the rainfall had been low that particular year? Of course not; according to the assumption given, there was no such year.

In the same way, light which began traveling ten thousand years ago cannot be said to have properties it *would* have had, had it originated at the star a million years ago. If God has revealed the time of creation, the thin tree ring does not indicate a dry year and the light does not indicate a million-light-year-distant star. In neither case should it be maintained that God is faithful only if the dry year, or the distant star, exists.

These ideas are presented here to enable those persons who accept the idea of a very recent creation to be more consistent. If one believes that creation occurred only ten thousand years ago, he has no way of knowing of the existence of most of the supposed Milky Way Galaxy, of which we are a part, and which is supposed to be saucer-like and 80,000 light years in diameter. In addition, he could know of no other galaxy. If the boundary of the universe of which the astronomers speak is no farther away than the farthest supposedly-known galaxies, the universe of the astronomer is one hundred million billion times larger than the universe of one who believes nothing is known beyond a distance of ten thousand light years.

Unfortunately, many people believe that God tells us creation occurred only thousands of years ago and that there are billions of galaxies spread over a distance of billions of light years. One who holds to these two concepts should understand that they are not compatible. The ideas developed in the preceding chapter may enable him to understand that God has not given man a date of creation of only thousands of years ago. These Biblical ideas do not prove that some stars are millions or billions of light years distant and at least as many years old, but they do remove the inconsistency.

B. *Astronomical Methods.* A rough outline of the methods the astronomers use will be given. This description is not to be taken as an attempt to prove the astronomers are correct. A critical discussion of their work should be found only in the technical literature, where the discussants must be able to understand the nature of the complicated methods of the astronomer. Consequently, the only methods described here are those considered valid by all astronomers. Wherever there is broad agreement within a well-developed discipline, such as astronomy, there is virtual certainty that the conclusions agreed upon are correct (cf. Ch. 6). The minimum age of creation as determined by astronomical methods can be obtained by accepting that which is universally accepted among astronomers.

The astronomer develops his knowledge of the universe in a step-wise fashion. Thus, he makes conclusions about planets, stars, and galaxies of stars by making deductions first of all about nearby heavenly bodies, then about those farther and farther away. If the astronomer were allowed to make but one observation of a distant galaxy, he could determine neither its distance nor very much else about it. Before he can put his observation in a proper context, he needs information about galaxies and individual stars close to us.

The astronomer must determine at each step the *distance* of the heavenly body from the earth. When he knows an approximate distance, he can determine the size and the mass of the heavenly body, as well as many other facts concerning it. The first step is to determine distances within the solar system. These distances are large according to earth standards — the planets are millions of miles apart — but they are very small when compared with distances between stars.

The method used to determine solar system distances can be explained by using an illustration. If on a cloudless night an observer on earth sees the lights of an airplane, he can place them against the background of stars, a background which seems to be fixed. With only this information, he is not able to determine the altitude of the plane. If there is a second observer a mile away who views the plane at the same instant, he also sees the lights of the plane against the fixed background of stars, but to him the lights do not seem to be in the same position with respect to the fixed star pattern as they seem to the first observer. If the two observers know how far apart they are from each other, and if they compare their observations made at the same instant, they can deduce from simple geometry how high the plane is flying. (The information with which they begin consists of the angles and the included side of a triangle. With this information, the altitude of the triangle, which is the altitude of the plane, is easily calculated.)

In the same way, if a planet millions of miles away is viewed against the background of very distant stars by two observers on earth, and if the observers are only a mile apart, they see the planet against almost identical star backgrounds. Therefore, they cannot determine the planet's distance. However, if the observers are thousands of miles apart on the surface of the earth, the star background differs enough so that they can ascertain the distance of the planet.

By careful observation of the paths of planets around the sun, their relative distances from the sun can be ascertained. For example, the planet Pluto is forty times as far from the sun as is the earth. Thus, knowing one distance within the solar system enables the astronomer to know all distances within the solar system. The earth can then be known to be 93 million miles from the sun, Pluto 3.7 billion miles, etc. Calculating distances within the solar system constitutes the first step in determining other astronomical distances.

In the second step it is desired to determine distances to stars outside the solar system. The distance to the closest stars outside the solar system is thousands of times greater than any distance within the solar system. Therefore, no two observation points on earth are far enough apart to enable the observers to discern noticeably different star backgrounds for the closest stars. However, distances within the solar system are known, and so the distance between observation points much farther apart than any two points on earth can be known. As the earth travels in a near-circular orbit around the sun each year, it moves from a certain position to a position 186 million miles away (across the circle whose radius is the sun-to-earth distance of 93 million miles) six months later. Thus, two observations of the same star taken six months apart are taken from observation points 186 million miles apart. These observations, made of the star against a distant, fixed-star background, can be used to obtain the distance of nearby stars in the same way solar system distances were obtained in step one. In this way star distances up to 300 light years have been determined. These are truly great distances, as is easily seen when it is recalled that light travels the distance from the sun to the earth in eight minutes; in other words, the sun is eight light minutes from the earth.

To ascertain star distances greater than 300 light years, most methods require a determination of the so-called "absolute" brightness of a star or galaxy. "Absolute" brightness refers to the actual brightness of a star; "observed" brightness refers to how bright the star appears to an observer on earth. For example, if at night one sees the lights of a ship at sea which he knows uses 1000-watt lights radiating equally well in all directions, it is possible (the detailed explanation is not given here) to determine from the observed brightness the distance of the ship. Knowing the ship uses 1000-watt lights is the same as knowing the absolute brightness of the lights.

If an observer sees several ships known to use lights of the same brightness, and is able to determine the distances of a few close ships by some method such as the method using two observation points, he can then calculate the distance of more distant ships having lights of the same brightness.

Similarly, the astronomer with his telescope can see millions of stars and galaxies. His problem is to convert the brightnesses he observes into absolute brightnesses. Once he knows an absolute brightness, he can calculate the distance of a star or galaxy.

In this third step various methods of determining absolute brightness are utilized. Within the limit of 300 light years the distances of stars have been calculated by observations from two points. Certain properties of these stars can be related to their brightnesses. For example, the astronomer observes that certain stars vary in brightness. The manner in which the brightness varies with time is directly related to absolute brightness. Thus, the variation in observed brightness can be used to determine the absolute brightness of stars beyond 300 light years distant.

In another method, it has been observed that some stars possess a certain kind of blue color. The absolute brightness of all the nearby stars of this kind is the same. Therefore, it is reasonable to assume that wherever such a star is seen, its absolute brightness corresponds to that of those whose brightness is known. Just as with the distant ships in the illustration, it is possible to determine the distances of stars whose brightness is variable, or stars which are of the blue kind mentioned, even though these stars are at a distance of greater than 300 light years.

In addition, the astronomer determines the distances of certain kinds of clusters of stars by using the blue star distances just discussed. He notes that all these *clusters* are of the same absolute brightness. Therefore, whenever he identifies such a cluster, he knows its absolute brightness and, combining this with its observed brightness, he can calculate its distance. The three methods of determining absolute brightness, using variable brightness, blue stars, and clusters, overlap and reinforce each other so that distances of stars and galaxies up to 20 million light years can be determined with confidence.

For the fourth step, the properties of whole galaxies (a galaxy such as our own is an aggregate of about 100 billion stars) within the 20 million light year distance are first observed. It has been observed that the spiral galaxies are of about the same absolute brightness. It has also been discovered that the farther away a galaxy is, the more rapidly it is moving away from us. The absolute brightness of spirals and the speed of galaxy recession can each be used to determine galaxy distance *beyond* 20 million light years distant, up to a distance of billions of light years.

This is of necessity a sketchy presentation of how astronomical distances are determined. A more complete demonstration could include a discussion of the independent ways to proceed from step to step in the determination of greater and greater star distances. The independent methods agree. The most distant stars yet observed appear to have emitted light billions of years ago. There is no conceivable set of human errors which if corrected could change "billions" to "thousands." This astronomical evidence indicates that God's first creative acts occurred no later than several billion years ago.

Radioactivity and Age

Another method of determining age depends upon the age of rocks. The only rock-dating methods which will be considered here are those involving the transmutation of one element into another.

All matter is composed of one or more of the "elements." Iron is an example of an element. A piece of iron or of any other element can be divided and subdivided until individual atoms are obtained. Atoms are the smallest particles which possess the properties of the element. If the atom itself is broken down, the characteristic properties of the element are lost. There can be slight differences between the atoms of a given element. Each kind of atom for a given element is an "isotope" of that element. For example, there are three isotopes of the simplest element, hydrogen.

Certain atoms are "naturally" radioactive. Such atoms "spontaneously" change, usually forming another isotope of the same element or an isotope of a different element, with the emission of one or more particles.

For an aggregation of atoms of a given isotope, the frequency of these atomic "explosions," as one after another of the atoms changes, is easily measured with modern instruments. Thus, for a sample of uranium-238, one of the isotopes of the element uranium, the individual explosions can be observed and counted. In this sample there will be many explosions per second. From a knowledge of the number of explosions per second and the number of atoms in the uranium-238 sample (also easily determined), it can be calculated that one-half the atoms in the sample will change in 4.4 billion years.

When uranium-238 changes, the new isotope is also radioactive, changing into still another isotope, until after several steps a non-radioactive isotope, lead-206, is formed. In 4.51 billion years half the original uranium-238 atoms will have changed into lead-206. If all the lead-206 in a certain rock were derived from uranium-238 which was once in that rock, and if the rock neither gained nor lost either uranium-238 or lead-206 since the rock formed, how long the process has been going on could be determined from a knowledge of the amounts of those two substances now present. Depending upon the relative amounts of uranium-238 and lead-206 present, a rock age of billions of years might be determined.

In rocks there are several other pairs of isotopes similarly related. In 710 million years one-half the uranium-235 (not to be confused with uranium-238) atoms becomes, after several steps, lead-207; in 13.9 billion years, one-half the thorium-232 atoms becomes lead-208; in 1.31 billion years, one-half the potassium-40 becomes argon-40 and calcium-40 atoms.

From a determination of the amounts present of each isotope of a pair, one can deduce the age of the rock *if he knows that none of the end product was present initially*. There are ways of determining whether or not this condition is met. For example, the age of a single rock may be determined by the use of several different pairs of isotopes. Since the rate of change is different for each pair, the age determinations will not agree if any end product was present initially. If the independent ages agree, there is strong indication that none of the end product was present initially, and that the rock is indeed that old. In one instance, a certain rock found in Southern Rhodesia has been dated by means of the three isotope pairs in which lead is the end product, and an analysis of the data for each pair gives an age of about 2.8 billion years. Presumably this rock was "born" when molten material crystallized to form rock. With such a rock, a possible cross-check involves measuring the amount of helium, the by-product formed when each of these three lead isotopes appears. The same kind of cross-checking has been carried out many times with different rocks. Over and over again ages of millions or billions of years (i.e., the time which has elapsed since the rock of interest crystallized) have been found.

Has water or some other agent removed one of the isotopes to give a false age? There are checks for this also. For example, the abilities of uranium and thorium compounds to dissolve in water are different. Dissolution of a uranium or a thorium compound can therefore be detected, since the ages deduced from a consideration of the data for the various pairs would then not agree. In addition, if water or some other agent removes lead, or causes argon (a gas which appears as potassium changes) to leave, the age would be in error, but the erroneous age would be too *small*, not too large.

Could the rate of radioactive decay have changed because there were once different conditions, thereby leading to an incorrect age calculation? The tendency of a rate of change as conditions are changed can be determined. Consequently, it is known that even under the most extreme conditions of pressure and temperature, radioactivity rates change at the most only slightly. The changes are so slight that they could not be the source of a significant error in the age determinations.

Did the radioactivity rates change because natural law was different at one time, rather than because of conditions? This possibility can be ruled out because the law and the particle, the atom, are intimately bound up together. (Cf. Ch. 4 for a discussion of natural law, and of this particular problem.)

There are other good methods of determining rock age using the concept of radioactive decay, not described here. All the evidence taken together is overwhelming, and there seems to be virtually no doubt that some rocks are millions, and that others are billions, of years old.

Questions about the Scientific Evidence

Several objections to the scientific ideas presented in this chapter, made by those who hold to the idea that creation occurred only a few thousand years ago, can now be discussed.

A. *The Apparent Age Problem.* Could God have created the universe with an apparent age? God certainly could have created rocks with a certain uranium-thorium-lead-helium-potassium-calcium-argon ratio. He could have created not merely a few rocks with ratios pointing to millions or billions of years, but since very many have been examined, he could have created all of those which have been examined in that way. Without doubt he could have created light on its way to the earth from not just one but from a countless number of stars which have been observed. If God did these things, the Christian should admit freely that God created a large number of clues, all pointing to about the same conclusion. The many independently-determined conclusions agree. The different methods *do not* give random answers, with some indicating a very young age of the earth and stars, others a great age, and still others an intermediate age.

If God created the universe with a built-in, apparently great age, no *correct* scientific evidence could indicate anything other than a great age. If one believes God created the universe with an apparently great age, he should not attempt to prove that the evidence indicating great age is weak. God would not create *weak* evidence. It is very difficult to understand why it is sometimes maintained that God created the universe with great apparent age, while the dating methods which indicate great age are attacked.

The apparent-age questions must be evaluated theologically, not scientifically. Scoffing at the scientific clues cannot possibly solve this theological question. The only proper way to answer this theological question is to assume temporarily, for the sake of the investigation, that the clues do indeed point to a great age. Then it may be possible to decide on the basis of what the Bible teaches whether or not God would create clues pointing away from the truth. It *seems* that God would not create in such a way; but the best answer that man can obtain to this question must be the answer developed carefully by theologians who are faithful to the Bible.

B. *Uniformitarianism.* Does not this use of scientific evidence assume without proof the principle of uniformitarianism? The Christian position is that God's creative activity ceased at the end of the six days of creation. Whenever uranium first appeared, its rate of decay was established. Bible believers must accept this example of uniformitarianism. There are applications of the uniformitarian principle to methods of dating which have not been discussed, and what is said here is not to be considered an automatic endorsement of all such applications. The best methods of determining age, the ones which have been described, seem not to be vulnerable. Natural law cannot have changed since the creation week, and therefore the methods are valid.

In the present discussion it is enough to note that the radioactivity decay rate of a certain isotope is one of the necessary attributes of the isotope. (A similar claim could be made for the various laws of astronomy which have been used.) This decay rate is as closely related to the nature of the isotope as, for example, its mass; one could never correctly maintain that the isotope once had no mass, or even a different mass. Were one to make such an absurd claim, he might as well maintain that at one time some other property of a substance – for example, its yellowness – was not the same property. The atom is the sum of its properties, and its radioactivity decay rate is one of those properties.

It is theoretically possible that certain laws used in this discussion change with time. If so, the correct law, the one which is inherently a part of the object studied, is the law which includes the time factor. Such a consideration does not affect the great-age conclusion. The amount of law non-uniformity needed to change the determined ages from millions or billions of years to thousands of years, is far greater than the amount of non-uniformity which could possibly exist undetected.

C. *Variable Scientific Conclusions.* Since scientists occasionally change their estimate of the age of the earth by as much as a billion years, can any trust be put in their current conclusions?

A change of a billion years in the estimated age seems at first like a very great change indeed. If this is a change from three billion to four billion years, the error is no greater than that made by a person who estimates a child to be three years old when the child is actually four. The length of the year as a unit of measure is purely relative. If instead of the year a hundredth of a second is the time unit used, a four-year-old child and a three-year-old child differ in age by 3.2 billion of these hundredth-of-a-second units. The scientist is quite accustomed to using very large and very small numbers, and he is always surprised when a non-scientist is skeptical of a scientific conclusion because it involves very large or very small numbers.

None of the revisions in the estimated ages of the earth, other planets, and stars affects the idea that these ages are great. Scientists are at present attempting to decide if the age of the earth is closer to four billion than to five billion years. This estimated age is not about to be changed to thousands of years.

D. *The Creation of Light and the Creation of the Sun.* The Bible teaches that the sun, moon, and stars were created on the fourth day. Therefore, is it not sinful to assume that these heavenly bodies existed during the events of the first three days? If the creation days were long periods, the creation of the sun, moon, and stars on the fourth day is indeed a problem. It is not always realized that the question of the fourth day is also a problem for those who hold to creation in six solar-length days. The problem which arises if solar-length days are assumed should be discussed first.

If the first three days were each about twenty-four hours in length, what was the source of the light which was created on the first day? It has been suggested that since God is light, he could have illumined with this light. When, however, it is said that God is light, reference is made to an attribute of God. God and his attributes are uncreated, and therefore this is not the light which was created on the first day.

Another suggestion is that God could create light without a light-bearer. Naturally, he could create in such a way. One should, however, give careful attention to what the Bible states concerning light on these first three days. On the first day light appeared at God's command, and there was light and darkness during the day (referring, of course, to the day of solar length) and night, respectively (Gen. 1:4-5). God created light, or its source, on the first day. Did he re-create it on the second day and again on the third – since, after all, darkness did follow each daylight period? But the re-creation suggestion is surely absurd. Such a suggestion is clearly an unwarranted, and therefore improper, addition to the Bible. The alternative, still assuming these three days were solar-length days, is that the light, or its source, created on the first day continued to exist, but that darkness appeared because the earth rotated. That is the same reason darkness appears today.

The appearance of darkness because of the rotation of the earth, on the days before the fourth day, is most important. First, it is thereby established that the point of view Moses is presenting (and this holds for either the solar length day or the long period day idea) is that of an observer on the surface of the earth. For, in interplanetary space there is no alternation between light and dark. Second, the appearance of darkness, even though light did not vanish from creation each day, means that the source of the light created on the first day was *localized.* Light did not appear everywhere; otherwise, the hypothetical observer on the surface of the earth would not experience darkness after a time of light. It is thus shown that the light created on the first day continued, that its source was outside the earth, and that it was localized. Furthermore, the account of the second and third days indicates that the intensity of this light was such that water and plants existed on the surface of the earth. If the earth were now either appreciably warmer or appreciably colder than it is, neither water nor plants could exist. Thus, we know fairly well the intensity of the light which illuminated the earth during the first three days. *In other words, the Bible describes in the account of these three creation days a light source which is just like the sun* – continuous, localized, and of the sun's intensity.

Therefore, the account of the creation of the sun, moon, and stars on the fourth day (Gen. 1:14-19) does not imply that there were no heavenly bodies earlier than that day. It has just been shown that there was one heavenly body – a body very similar to the sun if it was not the sun itself – which existed from the first day onward. It is therefore possible that the account of the fourth day teaches that on this day these distinct heavenly bodies were visible for the first time from the surface of the earth.

On the first creation day the hypothetical observer on the surface of the earth may have seen light, but not distinct heavenly bodies. He would have observed night as well. On the fourth creation day this observer would have seen the sun, moon, and stars distinctly for the first time.

Up to this point in the discussion of the creation of light, solar length creation days have been assumed. Using *this* assumption, it seems not to be correct to maintain that the sun, moon, and stars did not exist before the fourth day, and it is possible that all heavenly bodies and the earth were created on the first day. These conclusions can also be made if the creation days were long periods, assuming here also that light and dark alternated in solar length days. In fact, postulating creation days to be long periods enables one to explain rather simply why distinct heavenly bodies appeared only some time after they provided light.[1] It is possible that the early earth atmosphere was largely carbon dioxide, devoid of oxygen. (There is abundant mineralogical evidence that the atmosphere once contained very much carbon dioxide instead of oxygen.) A carbon dioxide "blanket" would have kept the earth warmer on the first day and the first part of the second day than it is now. Even today, a slight increase in atmospheric temperature is possible as large-scale combustion of fuels causes a very small carbon dioxide enrichment of the atmosphere. With a carbon dioxide blanket, much of the water present would remain as water vapor, and the weather would be continually cloudy. When the plants were created, they consumed carbon dioxide and produced oxygen. As the plants covered the earth, the atmosphere would have been converted over a long period of time from one containing very much carbon dioxide to one containing as much oxygen as it does contain. The temperature would decrease, the cloudiness would largely disappear because of condensation, and on the fourth creation day, the sun, moon, and stars would be visible from the surface of the earth.

Summarizing, adherents of either the solar length creation days or the period concept must consider carefully the meaning of the Biblical description of the first four days. Even if the creation days are assumed to be of solar length, there are certain inevitable deductions from the Biblical description which are not, however, always realized by those who accept the concept of creation days of solar length. These same deductions are consistent with the period concept. Using this concept, it is possible to provide at least one explanation of what the Bible teaches concerning the first four days.

E. *Carbon-14.* Has not the carbon-14 method of dating been shown to be in error, indicating that the radioactivity methods for determining the age of the earth are invalid? Carbon-14 has absolutely nothing to do with determining the age of the earth. With the carbon-14 isotope, half the atoms change in 5,730 years, half of what remains change in the next 5,730 years, etc., and so nothing older than about 40,000 years can be dated using carbon-14. Thus, for determining ages in the range of millions or billions of years, carbon-14 has not been used and it cannot be used. For some strange reason, one of the most common attacks on the idea of great age has been an attack on the quite irrelevant carbon-14 dating method.

Interestingly, the carbon-14 method of dating objects only a few thousand years old is both reliable and of use in refuting attacks made on the trustworthiness of the text of the Bible. For example, the carbon-14 dating of the Dead Sea scrolls showed that we now have a text of Isaiah about a millennium older than the previous text. Consequently, it is known that the previous text is an unbelievably faithful copy.

F. *Catastrophism.* In the use of dating methods, is proper account taken of catastrophes, such as the Noahic flood, which might account for evidence which falsely leads us to conclude great age? In this discussion there has been a deliberate avoidance of the use of dating methods which might be susceptible to the usual kind of catastrophe cited. The flood had no effect on the rates of radioactive decay. Neither the pressure caused by the water nor any other action of the water could have any effect on the very reliable radioactivity clock. Similarly, the astronomical evidence is not affected by any of these catastrophes. We are receiving light today which left stars billions of years *before* any of the postulated catastrophes occurred.

The relation between catastrophes and the dating methods not considered in the present discussion is a complicated, technical matter. Since these dating methods are not needed to establish the great age of creation, neither the methods nor criticisms of the methods are relevant here.

In Retrospect

Consider what the picture presented in this chapter means. Even though the essence of the miraculous is not the speed at which God acts, is it not possible that a picture of creation involving both vast time and space is more impressive than one with short time and relatively small space? God does not work using man's clock. If the universe is several billion years old, even this is a finite length of time, and it is as nothing compared to the eternal existence of God. In the same way, a finite universe with dimensions of billions of light years is as nothing compared to the infinite God.

God teaches that animals and plants reproduce after their kinds. It is one thing if this promise of continuity has been demonstrated for a period of thousands of years. It is quite another matter if the continuity has existed for hundreds of millions of years. Perhaps this picture of the universe is a picture so magnificent that we are moved even more than before to praise our creator God.

REFERENCES

1. Tanner, W. F., *Journal of the American Scientific Affiliation, 16(3),* 82 (September, 1964).

11 *The Origin of the Simplest Life*

The third evolutionary proposition, given earlier, is, "Life in the form of one-celled organisms evolved from non-living matter." The proposition is related to the other propositions, particularly the fourth proposition, concerning evolution of one kind of life from another. But, the third proposition differs enough from the others to warrant a separate discussion.

Comparing the Third Proposition with the Others

Is the idea of life evolving from non-living matter basically different from the idea of one living organism evolving from another? In order to answer this question, it is necessary to answer two prior questions.

First, how did one-celled organisms supposedly evolve from non-living matter and more complicated organisms evolve from simpler forms of life? Certain chemical reactions, not necessarily the same in the two processes, are believed by evolutionists to have taken place. The evolutionists might maintain that making a distinction between the two is artificial.

The second question is, "What is life?" Living material contains atoms, ions, and molecules, each of which is a non-living substance. There is nothing in living matter which is not accounted for: there is no mysterious "vital force," if "vital force" refers to a still-hidden force which conceivably could be traced to matter and identified by sufficiently careful analysis of the living substance. The connection between mind and matter is not relevant here.

In spite of these considerations, scientists have attempted to differentiate between living and non-living matter. They attempt to formulate criteria whereby one may distinguish between life and non-life. Without entering into a discussion of these criteria, it is sufficient to observe that even the components of the nucleus of the living cell are vastly more complex than any material not associated with living organisms.

Is evolution from non-life to life different from evolution of life from life? For the natural man – who denies the idea of creation by holding that man descended from animals, that complex animals and plants evolved from the simplest form of life, that the simplest form of life evolved from non-living matter, and that matter itself is part of an eternal universe – there is no break in the evolutionary chain. For the natural man, it is arbitrary to separate evolution of non-living to living from evolution of one kind of life to another. For him there is no qualitative difference between postulating chemical reactions which caused the evolution of life from non-life, and postulating certain other chemical reactions which caused evolution of one living organism from another.

In order to continue the discussion properly, it must be considered that separation of the two kinds of evolution is made arbitrarily for practical reasons. Examination of this proposition requires an analysis of some Biblical and scientific questions which are not discussed in connection with the other propositions. There is always the possibility that one might accept one kind of evolution, but not both. One might, for example, believe that God created the simplest form of life in a special creation act, but that subsequently there was organism-to-organism evolution.

The modern idea of the evolution of life from non-life is not the same as the "spontaneous generation" idea held a few centuries ago. "Spontaneous generation" means that suddenly, perhaps within hours or days, complex animals, such as insects or mice, spontaneously appear where only non-living substances, such as dirty rags, were present earlier. The mechanism of this change was assumed to be a mystery. Today, evolution of non-life to life supposedly occurs by means of a large number of successive chemical reactions. Simple, non-living substances, such as water and ammonia, are thought to be transformed eventually into the very complicated substances which make up the nucleus and the rest of the biological cell. These chemical reactions are in principle knowable. In fact, different possibilities for the reaction sequence have been suggested. The mechanism is an essential part of the modern theory, and there is no similarity to the older rags-to-mice spontaneous generation theory.

The Bible and the Origin of Life

Since this proposition is limited to the evolution of life from non-life, many of the Biblical passages which are usually cited in discussing evolution cannot be used. For example, when the Bible speaks of reproduction "after its kind," it seems fair to assume that only living things are being discussed. What can be said about the postulated change from the non-living to the living?

In Genesis 1 there is no direct mention of one-celled organisms, or of substances simpler than one-celled organisms, such as viruses and bacteriophages, also considered to be living. (In this discussion, one-celled organisms are arbitrarily taken to be the end-product of non-life-to-life evolution because the problem of producing the more complex cells from viruses or bacteriophages would still remain were the simpler substances considered to be the end product in non-life-to-life evolution.) In the creation account of Genesis 1 there is mentioned, in order, the creation of grass, herbs yielding seed, fruit trees, the moving creature that has life (in the water), fowl, whales, cattle, creeping things, beasts of the earth, and man. But this list is presented in the Bible as the list of living things which were created during creation "week," the "week" in which all creation took place. Thus, one-celled organisms must lie within one or more of the groups mentioned in the list. Perhaps the phrases "moving creature," "creeping thing," and "beast of the earth" are meant to include these organisms.

Is it possible that these Genesis 1 phrases include one-celled organisms, since, after all, Moses and the Israelites who first read Genesis did not know of such organisms? There are undoubtedly many plants and animals about which these Israelites knew nothing. It cannot be correct to rule out of the creation account all these living things. In the same way, "He made the stars also" (Gen. 1:16) refers to innumerable stars and galaxies about which these Israelites knew nothing. Israelite ignorance of the existence of one-celled organisms does not mean that their creation is not indicated in Genesis 1.

Although it can be assumed that the creation of one-celled organisms is included in the Genesis account, mere inclusion says nothing about the truth or falsity of Proposition 3. There are still two key questions to be answered.

A. *Creation from Nothing.* Did God create one-celled organisms from nothing, or did he use pre-existing material? Creation from nothing is nowhere explicitly mentioned in the creation account. Yet, it is obvious – since these events constituted the beginning of the history of creation – that ultimately there was creation from nothing. From a logical point of view it is not necessarily true that each thing was made separately from nothing. There is the possibility, unless the Bible teaches otherwise, that certain things were derived from others. In this connection, the meanings of the two Hebrew words used to describe God's creation of animals are of interest. Other Biblical uses of these words indicate that neither word necessarily teaches that animals were created from nothing.

(1) The first word is *bara,* the Hebrew word for "created" in the following creation passage:

> And God created great whales, and every living thing that moveth, which the waters brought forth abundantly, after their kind, and every winged fowl after his kind: and God saw that it was good. (Gen. 1:21)

In the Bible *bara* is usually associated with an activity of God. Examples of such an activity are the works of the creation week, such as

> For lo, he that formeth the mountains, and createth the wind. . . .(Amos 4:13)

or other activities uniquely divine, as

> For behold, I create new heavens and a new earth. . . . (Isa. 65:18)
> Create in me a clean heart, O God. . . . (Ps. 51:10)

> This shall be written for the generations to come: and the people which shall be created shall praise the Lord. (Ps. 102:18)

These five passages are typical of the passages in which *bara* denotes an activity of God. Is it possible, from an examination of the last four passages, to decide whether or not *bara* in Genesis 1:21, where it refers to the creation of "every living thing that moveth," means creation from nothing?

When *bara* is used in "I create new heavens and a new earth," creation of at least some aspect of the new heavens and earth from nothing seems to be indicated. "Formeth the mountains, and createth the wind" seems to imply both creation from nothing and later developmental activity of God. What of "Create in me a clean heart"? Cleanliness comes from without as the Spirit applies Christ's work, but here too a developmental aspect is implied. In the last passage quoted, the Bible speaks of "the people which shall be created." It would probably be presumptuous to assume that human birth is creation from nothing, and that development is not involved. Thus, even when *bara* refers to an activity of God, creation from nothing may or may not be indicated.

(2) The second Hebrew word used to denote the idea of the creation of animals is *asah*, translated as "made" in the following passages:

> And God made the beast of the earth after his kind. . . . (Gen. 1:25)
>
> Now the serpent was more subtle than any beast of the field which the Lord God had made. (Gen. 3:1)

It is instructive to observe how *asah* is used in other passages. Following are some typical uses, where *asah* is translated "made" or "make":

> And God made the firmament. . . . (Gen. 1:7)
>
> And God said, Let us make man in our image, after our likeness. . . . (Gen. 1:26)
>
> . . . In the day that the Lord God made the earth and the heavens. (Gen. 2:4)
>
> . . . And they sewed fig leaves together, and made themselves aprons. (Gen. 3:7)
>
> Make thee an ark of gopher wood. . . . (Gen. 6:14)
>
> Now Israel loved Joseph more than all his children, because he was the son of his old age, and he made him a coat of many colors. (Gen. 37:3)

The conclusion to be made concerning *asah* is similar to the one made concerning *bara*. *Asah* can denote creation from nothing or other activity which is uniquely divine, but it can also denote human activity which has no relation to the creation-from-nothing concept. Thus, in answer to the question concerning whether or not God created the simplest life from nothing, it must be stated that there is no *certain* Biblical information. Sometimes *bara* and *asah*, in describing God's activities, clearly mean creation from nothing. At other times, there was pre-existing material. For example, man was made *(asah)* (Gen. 1:26) from the dust of the ground. It is not as certain that there can be pre-existing material when *bara* refers to divine activity. Examination of the passages cited suggests caution in making a dogmatic statement.

B. *Natural or Miraculous Origin.* If the simplest life was derived from the inanimate, could such a transformation have been natural, or would it have been of necessity a miraculous transformation? Natural law is but human correlation; all events are ultimately caused by God, and miraculous events are those which fit into God's law but not into man's correlation (cf. Ch. 4). With this understanding of "natural" and "miraculous" events, it may properly be asked what occurred when the simplest life first appeared. Was there a sequence of chemical reactions, similar to one of the sequences proposed by evolutionary biochemists, a sequence which could be termed "natural"? Or, was there a miracle (such as the miraculous appearance of manna to the children of Israel in the wilderness) performed as the simplest life first appeared?

The Bible does not answer directly the question concerning whether or not the first appearance of life was miraculous. Nor does it answer the question concerning whether or not life arose from nothing. Is there difficulty in finding Biblical answers because the wrong questions have been asked? It has often been claimed that asking these scientific questions of the Bible is very much out of place. But, as was shown in earlier chapters, the Bible does indeed give us some information which sheds light on profound scientific questions. Why does the Bible not help us in *this* most fundamental question concerning the origin of the simplest life? God could have told us in the Bible that he used rocks, water, air — anything to communicate the idea of inanimate matter — to make life, or that he made life from nothing, without using pre-existing matter. The Bible may be silent on this matter in order that man can deduce from his science that God was active in a very special way in creation. Man has used his science to supplement what he learns from the Bible; man has, in fact, already used his science to show how ridiculous is the old idea of spontaneous generation. The next section suggests why the Bible does not answer directly these questions concerning the origin of the simplest life.

Science and the Origin of Life

In recent years biologists and biochemists have been very interested in the origin of the simplest life, assuming that living matter evolved from non-living matter. They have been interested in *how* such a transformation occurred, not in *whether* it occurred. One might well ask what processes, whereby life arose from non-life, are being considered.

A. I. Oparin, a Russian biochemist, led the way in 1938 in his *Origin of Life* in describing what must have occurred as the simplest life arose from non-life.[1] Oparin postulated that billions of years ago hydrocarbons, compounds of hydrogen and carbon, formed on the surface of the earth, and that the earth's atmosphere was steam. By chemical reaction ammonia was formed. These chemicals – the hydrocarbons, ammonia, and water in the form of steam – reacted, forming molecules containing carbon, hydrogen, oxygen, and nitrogen. These molecules were included in coagulated materials in the seas. Eventually, these molecules reacted to form proteins, and large numbers of these proteins combined to make the nucleus of the living cell. The material in the cell outside the nucleus, the cytoplasm, was also formed by a gradual process, and eventually the simplest organisms, single-celled organisms, appeared.

It is possible that the earth is billions of years old and that the primitive conditions Oparin describes, including the presence of steam, hydrocarbons, and ammonia, actually existed. There remains the key question: What is the probability that these complicated molecules actually formed and subsequently combined to make up living cells in the time available?

Non-Life to Life Probabilities

In order to discuss this question, a short explanation of a few probability principles is necessary.

A. *Some Probability Considerations.* If one flips a coin, the probability that it will be heads is one in two. If it is flipped again, the probability for heads is again one in two. However, if the outcome is considered before flipping the coin the first time, the probability of two consecutive heads is one in four. Similarly, the probability of three consecutive heads is one in eight; four, one in sixteen; and, carrying it much farther, ten, one in 1,024; twenty, one in 1,048,576. If not consecutive heads, but some other pattern is decided upon before the flipping begins, what is the probability one will flip, in this order, for example, three heads, five tails, one head, one tail, four heads, two tails, two heads, and two tails? The probability for *any* sequence of twenty flips decided upon before the flipping begins is the same, one in 1,048,576.

Scientists are accustomed to calculating probabilities for physical systems. For example, chemists and physicists frequently calculate how many times per second the gas molecules in a given vessel collide with the vessel's walls and with each other. Such a calculation is ultimately a probability calculation. In the question of the production of life from non-life, the probability that a postulated sequence of reactions did indeed take place in the time available, billions of years, has been calculated. The relevant questions are those concerned with the probability (1) that the molecules which contained carbon, hydrogen, oxygen, and nitrogen actually formed proteins, (2) that large numbers of these proteins associated together to form a single nucleus, and (3) that this nucleus combined with cytoplasm to make up the cell.

Since different assumptions concerning the precise primitive earth conditions can be made, the probabilities calculated can be expressed in different ways. These differences are not important, since the different calculations present the same general picture. Two such calculations, made by scientists who accept the possibility of living matter evolving from non-living matter, are cited here.

B. *Typical Non-Life to Life Probability Calculations.* (1) The protein components of the cell are in turn the product of the reaction of many amino acids. The amino acid molecules, containing hydrogen, oxygen, and nitrogen, may have existed on the primitive earth. Although amino acids will combine to form larger molecules under favorable conditions, it is unlikely that these larger molecules will be of biological interest. For the new molecule to be such a protein, there must be a certain orderly arrangement of the amino acids in the larger molecule. The question of amino acid order is one way in which the question of probability arises. Suppose a photographer taking a group picture of a school class of thirty children wishes to arrange the children alphabetically, even though he knows none of their names. If he arranges the thirty children without help, what is the probability that he will indeed arrange them in alphabetical order? Obviously, such a probability is exceedingly small. Even with many trials there is only a very small probability that he will arrange them correctly.

J. Lever, who maintains that living matter could have evolved from non-living matter, discusses the calculations of several investigators relevant to the amino acid-to-protein problem:

> What is the probability that from the free interplay of amino acid molecules there originates a protein built up of 1000 amino acids in a definite arrangement as they always occur in normal proteins? It appears that by considering a great number of simplifications this probability amounts to 10^{-1360} if for instance 1000 kg. [about one ton] of amino acids gets the chance for this reaction during a billion years. This means thus a decimal fraction with 1359 zeros after the decimal point.[2]

Consider how small a probability of 10^{-1360} is. In a cup of water there are about 10^{25} (i.e., 10 followed by twenty-four zeros) atoms. If one could remove from the cup *one* atom, the probability he would select a certain, predesignated atom is 10^{-25}. Considering all the atoms of the earth, which contains about 10^{50} atoms, the probability of selecting a predesignated atom is 10^{-50}. What about selecting one particle from the entire observed universe, with billions of galaxies, most of which contain 100 billion suns? The probability of selecting a certain, predesignated particle in the universe (always, of course, selecting at random) is about 10^{-80}. Imagine another kind of a universe, so large that *each* particle in our universe becomes as large a universe of particles as our universe. This hypothetical universe would contain 10^{80} times as many particles as our present universe. The probability of selecting at random a certain, predesignated particle in this hypothetical universe is 10^{-160}.

If that larger universe is again expanded in the same way – each particle begetting a new universe as large as this larger universe – the probability becomes 10^{-320}. With the next expansion, it is 10^{-640}; the next, 10^{-1280}. This is "near" the amino-acid-to-protein-in-a-billion-years probability of 10^{-1360}. Obviously, increasing the amount of amino acid from a ton to a fantastic amount increases the probability only insignificantly.

That is not all. A protein is not alive. Large numbers of proteins are needed to make up the cell nucleus. Furthermore, the nucleus needs the cytoplasm. When these substances are brought together in the right way, then and only then is there a living one-celled organism. The various components of the cell must be synthesized at the same time. The cytoplasm cannot continue to exist without the nucleus, and the nucleus cannot continue to exist without the cytoplasm. Finally, for these events to have any significance for the formation of life, there must have been produced a large number of cells, not just one cell.

These probabilities are so small that they show the spontaneous generation of life in the primitive earth to be a fantastically improbable event.

(2) Lecomte du Nouy, who believed that living matter evolved from non-living matter, expressed in a different way the probability that a protein molecule would form under primitive earth conditions.[3] He calculated that on the average, in a volume of protein components equal to the volume of the earth, *one* protein molecule would form *in 10^{243} billion years.* There have been objections that such a calculation improperly assumes a single jump from the components to the protein product.[4] There are few intermediate steps which are feasible stopping places, since if anything other than the protein product (or perhaps one of the few stable intermediate products) is made, there would be decomposition back to the starting materials. It may be possible to show that du Nouy's probability and the one Lever quotes should be larger; but it does not seem reasonable that any such change could be large enough to be meaningful. In any event, the formation of a protein molecule would be only a very small beginning step in the formation of the living cell.

If these calculations are correct, how is it that proteins are constantly being made in living systems? The components of the cell aid each other; the protein can be synthesized very rapidly when an "enzyme," a cell component, is present. An enzyme is a biological catalyst, that is, a biological material which speeds up certain chemical reactions. But making the enzyme without living material is just as difficult as making a protein without an enzyme.

Because of the complexity of cells, scientists are a long way from synthesizing them in the laboratory. Occasionally an announcement is made concerning a very small step in this direction, and news media convey the idea that man is near the big breakthrough – the synthesis of life from non-living material without the aid of living material. It is quite possible that man will never achieve such a synthesis. But even if he does, the guided laboratory experiment is not to be compared with the random motions of molecules in the primitive earth. Thus, it is easy to *arrange* twenty coins in any desired heads-tails order, but the chances of the coins being flipped in the desired order are vanishingly small.

Are the scientists and philosophers who accept the idea that the first cells arose spontaneously unaware of the low probability of such an event? In general, this low probability is both understood and accepted. The scientists and philosophers who today accept the idea of evolution of life from non-life do not disagree, in principle, with du Nouy's calculations of a generation ago (although his solution to the problem thus posed is not popular). It is of interest to determine why Oparin, who developed the idea that proteins arose from non-living matter, felt that the idea that life evolved from non-life should be accepted. Approving F. Engels' ideas on evolution, Oparin said,

> Engels shows that a consistent materialistic philosophy can follow only a single path in the attempt to solve the problem of the origin of life. Life has neither arisen spontaneously nor has it existed eternally. It must have, therefore, resulted from a long evolution of matter, its origin being merely one step in the course of its historical development.[5]

In one way or another, the argument used even today is expressed as follows: Life was not created instantaneously, "miraculously"; yet, life exists. At one time life obviously did not exist; therefore life *must* have evolved. The "must" in the Oparin quotation is probably the most important word in his book.

In the analogy of flipping a penny, *any* predesignated order of flipping twenty pennies is as improbable as any other order. It is not merely flipping twenty consecutive heads which is improbable. Similarly, the sequence of chemical events and atomic motion *which actually occurred* the first few billion years of earth history is also improbable if this sequence is considered *beforehand*. Thus, *any* sequence of chemical events is as improbable as one which would produce life. An improbable sequence which produced life is believed by the evolutionist to be the sequence which actually occurred. With this belief the Christian must either agree or disagree.

Probability, Creation, and Miracles

Did God use an improbable sequence of chemical events as the means of creating life? God acts either through natural law or through a miracle. The natural man does not admit the existence of either miracles or God. The natural man is therefore forced to accept Oparin's and Engels' dictum: "[life] *must* have . . . resulted from a long evolution of matter. . . ." (Italics added.) A rejection of miracles thus implies an acceptance of a long evolution of life from non-life. If life *must* have so evolved, it is because a miraculous creation *must* be impossible. One who accepts either of the following statements is thus forced to accept the other: (1) There *must* have been a long, natural evolution of life from non-life. (2) Miracles, contrary to natural law, *cannot* have occurred.

An attempt will now be made to show that if the second of these two statements is true, the first is not true. In other words, accepting the impossibility of miracles leads to a logical contradiction.

This contradiction is shown by considering a rather strange twist in the twentieth-century discussion of miracles. Up to the twentieth century, men always considered the miracles recorded in the Bible to be impossible if natural law is to be universally obeyed. There was always the possibility that a few miracles might be understood when scientific knowledge increased.

The situation changed with the advent of quantum mechanics in the twentieth century. Physicists learned that matter possesses heretofore undreamed-of properties. Particles cannot be exactly placed in space. Physicists speak of probability rather than certitude. (For a discussion of another consequence of this probability approach, see Ch. 5.) For example, there is a high *probability* that a given electron is within a certain small region at a given instant, but there is always the *possibility* it is at another place. A similar statement can be made for all particles in all atoms. As a means of teaching quantum mechanics to the beginning student, some bizarre possibilities have been suggested. In a famous tongue-in-cheek example, Gamow pointed out that it is *possible* that a car left in a locked garage one evening will be found outside the garage the following morning. The particles of which the car is comprised *could* do that which is extremely improbable. To the average nonscientist, such discussion seems – to put it mildly – irresponsible. Yet, physicists do not scoff at Gamow's reasoning. Oddities such as finding the car outside the garage are not observed because their occurrence is fantastically improbable. An extremely important point is that the probability that such an odd event will occur is at least in principle, and sometimes actually, calculable.

When these strange ideas were first presented, it was suggested that *here* is a way that science can prove that at least some of the miracles of the Bible are possible. We do not observe a metal ax-head floating on water. Yet, such an event is, acccording to quantum mechanics, not impossible. Although the prophet was the apparent cause of this Biblical miracle (II Kings 6:6), what actually occurred was thought to be improbable, but not impossible. God could operate through the "loophole" provided by quantum mechanics. The natural law of man which explained miracles could therefore begin to assume an absolute character. Perhaps God could eventually be left out of the process; then, miracles would be events which "just occurred," that is, chance events.

Scientists accept no part of this attempted explanation of miracles. The Christian scientist does not need such tortuous reasoning, since he realizes man's laws are not God's. The natural man who is a scientist knows it would be foolish to believe such an improbable event occurred. The natural man accepts two statements which are relevant to this question: (1) The probability of the ax-head floating, or of some other miracle occurring is, in principle, calculable; and any such probability is very, very small. (2) The basic reason miracles are rejected is not that the testimony concerning them is weak, but *miracles are rejected because they cannot occur* (cf. Ch. 4). The low quantum-mechanical probability that the ax-head floated indicates to the natural man that such an event *cannot* occur.

Consider once again the question of the probability of the evolution of life from non-life. Calculations indicate that the probability of such evolution is unbelievably small. The natural man is becoming more convinced that *this* improbable event occurred; it *must* have occurred, according to Oparin, since a "miraculous" creation is ruled out. Now, the point of this discussion is this: the improbable event the natural man accepts – the evolution of life from non-life – *is a miracle* in essentially the same way that the floating ax-head is a miracle. Probabilities can be shown in both cases to be so small that one ought not to be very interested in them. Accepting the second of the two statements given above, "Miracles, contrary to natural law, cannot have occurred," invalidates the first statement, "There must have been a long, natural evolution of life from non-life."

Is this argument invalid if the two probabilities are not equal, with the probability of the ax-head's floating much less than the probability of evolution of life from non-life? Suppose, for the sake of the discussion, that the probabilities do differ. The argument presented is still valid, since the natural man will not admit that the ax-head floated even if its floating is "only" as improbable as the evolution of life from non-life. If the natural man learns that the quantum-mechanical probability of the ax-head floating is such that this event would occur on the average once in 10^{243} billion years (du Nouy's probability for the synthesis of a single protein under the conditions given earlier), he would conclude that the observers who reported the event of II Kings 6:6 did not actually see the ax-head float.

The discussion concerning the natural man's inconsistency concerning the evolution of life from non-life may be summarized as follows:

(1) The natural man debunks the evidence for miracles, but his primary objection to miracles is that they could not have occurred.

(2) Therefore, according to the natural man, the origin of life is non-miraculous.

(3) The only remaining way in which life could have originated, according to the natural man, is by evolution of the simplest living material from non-living material.

(4) The probability of such evolution has been shown to be vanishingly small.

(5) Many miracles are not physically impossible, but the probability of their occurrence, calculated according to modern physical theory, is also vanishingly small.

(6) Accepting evolution despite its vanishingly small probability (point [4]) requires that the vanishingly small probability of miracles (point [5]) cannot be ruled out, even if it should be that the probability of miracles is smaller.

(7) The natural man rules out all miracles, even if the probability of a miracle could be shown to be no less than the probability of evolution of life. Therefore, the reasoning of the natural man is internally inconsistent. If he accepts evolution of life from non-life, he must admit the *possibility* of a miracle. On the other hand, if he holds that no miracle can seriously be considered to be explained by quantum mechanical reasoning, he should by the same token reject the non-life to life possibility. He cannot reject the non-life to life possibility, since the only alternative, as Oparin states, is something like a miracle, the very things he has denied.

Le Comte du Nouy also identified non-life to life evolution with the miraculous without making the comparison with Biblical miracles. He realized that denying the miraculous leads to an inconsistency. Even though he believed that evolution did occur, he said,

> Either we have absolute confidence in our science and in the mathematical and other reasonings which enable us to give a satisfactory explanation of the phenomena surrounding us — in which case we are forced to recognize that

> certain fundamental problems escape us and that their explanation amounts to admitting a *miracle* – or else we doubt the universality of our science and the possibility of explaining all natural phenomena by chance alone; and we fall back on a *miracle* or a *hyperscientific intervention.*[6] (Italics added)

Discussion of Objections

The objection that the probabilities for the occurrence of miracles and evolution may not be equal has been discussed. In addition, there are two other objections to the argument just presented.

A. *All Sequences Improbable.* Cannot non-life to life evolution be accepted on the basis that *any* sequence of chemical events during a billion-year period is highly improbable, and that the sequence which occurred happens to be the improbable one which led to life? This is frequently the reason given for accepting non-life to life evolution. In other words, probabilities are considered irrelevant. Rather, the existence of life is taken to be sufficient proof that an improbable sequence did occur. Consistency demands that no event be rejected because its occurrence is improbable. If the natural man accepts evolution on the basis that life has, after all, occurred, he should also examine the *evidence* for each miracle. Perhaps a miracle could then be accepted; at the very least, not *all* miracles should be rejected. The individual particles of all the atoms of the ax-head and the nearby water had to be *somewhere* – any one configuration would be very improbable – and so why not believe that the ax-head floated? Yet, the natural man will never examine the evidence for individual miracles. Miracles *as a class* are rejected. For this reason, he rejects the possibility of *future* miracles just as much as the possibility of miracles in the past. In the future, a miracle might occur and the evidence might be overwhelming – but, no matter, the evidence is not to be examined. For the natural man, miracles do not occur because they cannot occur.

The argument that some initially improbable sequence of events did indeed occur over a billion-year period, is also an argument for miracles; and once miracles are accepted, the original reason for insisting upon non-life to life evolution vanishes.

B. *Future Discoveries.* Is it not possible that new, presently undreamed-of scientific discoveries will reveal that the low probability of non-life to life evolution is wrong, and that this evolution was actually a highly probable event?

Such discoveries seem to be impossible. Even if this judgment is incorrect, the argument given in the previous section concerns the *present* attitude of the natural man. Without knowing that such discoveries are possible, he has accepted the idea that non-life to life evolution occurred. The same ideas concerning possible future discoveries could also be applied to miracles. The Bible believer does not rule out the possibility that at least some miracles will be explained. Some miracles may even now be explainable. For example, it is just possible that when Moses sweetened the waters at Marah by putting a tree in the water (Ex. 15:25), that he used an ion-exchange process now scientifically understood. If presently undreamed-of discoveries are postulated to solve the non-life to life riddle, why not postulate the same for the ax-head?

The Bible believer does not wait for scientific discoveries to prove that miracles occurred. It does not concern him if man's natural law never is able to explain miracles. Nor does the natural man wait; he rejects miracles until they can be shown to be probable events. The natural man accepts non-life to life evolution with the hope that the presently calculated probability will be shown to be much lower than the actual probability. At the same time he rejects miracles until the presently calculated probability of their occurrence can be shown to be incorrect. Postulating non-life to life evolution inevitably leads him into one more internal contradiction in his thinking.

What can the Christian say about the question of whether or not God used non-living matter to create the simplest life? In other words, what is the Christian answer to the third evolutionary proposition?

The possibility that God did indeed create the simplest life from non-living material has not been eliminated. Even though a whole world of amino acids would by random motion provide on the average only one protein in 10^{243} billion years, God could make a protein from its constituent amino acids in an instant. He could have used either slow or fast processes. He could have used processes which conform to natural law, or he could have used processes which do not and never will conform to man's law. He might have used previously existing matter to make the simplest life, or he might have created life from nothing. *But none of these methods is less than miraculous,* understanding "miraculous" to mean just what it usually means when it is used in connection with Biblical miracles.

Since the appearance of life was in any event a miracle, the question of whether or not pre-existing, non-living material was used is irrelevant. It was not essential to the performance of the miracle that Christ used water to make wine in Cana of Galilee (John 2:1-11). Nor were the clay and water important when he used them in healing a blind man's eyes (John 9:6-7). His feeding of the four thousand (Matt. 14:15-21) did not depend upon the small amount of fish and bread used. Whether or not the manna the Israelites received was created from nothing (Ex. 16:14-15) is of very little interest. In none of these instances are there theological debates concerning the role of the starting materials. These miracles of Jesus, the miracle of the manna, and many other Biblical miracles would seem very much the same were we to know that there were no starting materials.

Similarly, the question of starting materials for the creation of the simplest life is not of much importance. The Bible-believer knows, and the natural man should know from science, that irrespective of the question of starting materials, the appearance of life was a miracle.

A Possible Misconception

There have been attempts to disprove evolution which postulates order (a protein molecule) arising from disorder (randomly mixed molecules) by applying the second law of thermodynamics, which teaches that there is a universal tendency for order to produce disorder.

This argument is not valid because it is not a correct application of the second law. Order can very easily be produced from disorder. For example, when randomly moving water molecules in liquid water freeze to the ordered ice structure, order is produced from disorder. Many such examples could be given. The second law states that *in a closed system* order tends to produce disorder. The water system described is not a closed system. For water to freeze, it must lose heat. The heat lost is gained by something such as a piece of ice well below the freezing point of water. Whatever gains heat must be part of the system if the system is a closed system. Adding liquid water to such ice could cause liquid to freeze, as the temperature of the ice initially present would increase. Calculation of all the disorder changes involved, including the changes for both the ice which was ice at the beginning and for the water which became ice, would reveal that mixing the ice and the water brought about a net increase in disorder for the entire closed system. (The same result is obtained if the freezing of water in a refrigerator is considered, taking into account the power input, etc.)

If the second law of thermodynamics is not applied properly, one could err in an analysis of systems containing plants or animals. Consider the example of a growing plant. The plant is ordered; molecules, ions, and atoms are not arranged randomly. The plant can feed on gas molecules and substances dissolved in water near the plant roots; i.e., it can feed on disordered materials. It would seem that order is produced from disorder. But the plant with its nearby food is not a closed system. There is also energy input, usually in the form of light. The calculation of the change in order of the total closed system would be very complicated. There is no evidence that the second law does not hold for this system, even though at first one might suppose the second law is violated in living systems.

Just as plant growth cannot be shown to be contrary to the second law of thermodynamics, the idea of evolution of life from non-life cannot be shown to be contrary to this law. The two applications are parallel, since every statement which says that evolutionary theory postulates order arises from disorder can also be made about the growth of living things. With the postulated evolutionary process, energy and food are used by that which evolves. If the second law is assumed to be true, and *if* it is shown to be contrary to the evolution of life, then the second law could also be shown to lead to the ridiculous conclusion that there is no life. The second law can be assumed to be true, and no one who understands the closed-system condition has shown the law to be inconsistent with either the existence of life or the concept of evolution of life from non-life.

The second law has been applied in an attempt to refute not only non-life to life evolution, but also animal to animal and plant to plant evolution, since in these cases also evolution from the simple to the complex, with an increase in order, is postulated. Here too the "refutation" neglects the closed-system condition. The second law of thermodynamics seems to be worthless as far as any part of the evolution discussion is concerned, and it is not referred to further in this book.

Can Man Synthesize Life?

The complexity of the simplest life is so great compared to its non-living component parts, that the very appearance of the simplest life was a miracle. Can man use the same non-living materials to synthesize life? Can man work a miracle?

Not all Biblical miracles are necessarily that which man cannot achieve. If Moses sweetened the waters using the ion-exchange principle, Moses' act is still to be considered a miracle because it demonstrated God's overwhelming power. God could allow men to learn enough science to achieve some Biblical miracles.

The chemist has carried out many extremely complicated syntheses. Molecules once available only because they are synthesized by living things have indeed been made in the laboratory. From time to time it has been announced that some chemists have moved one more step towards synthesizing life in a test tube. However, such syntheses are accomplished by using materials *which are presently obtained only from living organisms,* even though such materials are themselves not alive. Obviously, such a laboratory synthesis is itself not analogous to the supposed primitive-earth synthesis of the *first* living material.

There is no doubt that in the laboratory the most complicated materials made without the use of the products of living matter are simple compared to any of the vast number of substances which must be made to produce, for example, a living cell. First, a very complex molecule must be made, and this synthesis must be repeated many, many times; and then these components must be assembled in precisely the correct way, even as the individual molecules must be assembled in just the correct way from their components. All this is so far from what can now be done, that there is hardly an analogy which can be used to express the difficulty.

When such a difficult problem is encountered, can it be said that man cannot *in principle* carry it out? Suppose it is maintained that there is no principial reason man cannot travel to a part of our galaxy which is fifty thousand light years away. Fantastic steps necessary for such an achievement could be conceived of. Man could learn to travel nearly as fast as light, he could put a colony on a space ship, and in a time a little greater than fifty thousand years he would arrive. It may be that the hope that life will be synthesized using only non-living materials is just as small as the hope of the would-be space travelers. The fantastic complexity of the problem seems to rule out its feasibility. But if God allows it, it is not *impossible.*

REFERENCES

1. Oparin, A. I., *Origin of Life* (tr. S. Morgulis), Macmillan, New York, New York, 1938.
2. Lever, J., *Creation and Evolution* (tr. P. Berkhout), Grand Rapids International Publications, Grand Rapids, 1958; p. 47.
3. Lecomte du Nouy, P., *Human Destiny*, David McKay Co., Inc., New York, New York, 1947; p. 34.
4. Hearn, W. R., and Hendry, R. A., "The Origin of Life" in *Evolution and Christian Thought Today* (R. Mixter, ed.), Wm. B. Eerdmans, Grand Rapids, Mich., 1959; p. 68.
5. Oparin, *op. cit.*, p. 33. Quotation from edition of Dover Publications, New York, 1953.
6. Du Nouy, *op. cit.*, p. 36.

12 Biological Evolution

The fourth evolutionary proposition is, "All animals and plants evolved from one-celled organisms." If the third proposition, concerning non-life to life evolution, is not true, unless it is understood to imply a miracle, then one might expect that neither was there a non-miraculous evolution beginning with one cell.

There are at least two reasons for discussing the fourth proposition even though a decision concerning the third proposition has been made. First, many evolutionists do not insist that all five propositions must be accepted. Sometimes the third proposition is rejected and the fourth is accepted. Why this pattern of rejection and acceptance exists is not of present concern. The other reason the fourth proposition merits discussion is that some of the Biblical and scientific considerations are of importance here whereas they have no bearing on the third proposition. For these same reasons, the discussion of the fifth proposition, concerning whether or not man descended from animals, is kept separate from the discussion of the other propositions. If one accepts or rejects each of the five propositions on its own merit, then his thinking on this subject will be more careful, scholarly, and honest.

What Is Meant by "Biological Evolution"

It is postulated that by a long series of small changes the simplest life (whether or not the simplest life is a single-celled organism is not relevant here) changed to more complex life until the entire spectrum of plants and animals was produced. This process is thought to be continuing, and that, given enough time, there will be many new plants and animals. The important questions are concerned with what these small changes are, and whether or not enough of them actually occurred to bring about the postulated result.

Occasionally there is minor variation from one generation to the next. Sometimes the variation is permanent, and it is this kind of variation, a "mutation," which is meaningful for evolutionary theory. For example, on a nineteenth-century New England farm a lamb with short, bowed legs was born. This lamb was the first of the Ancon breed of sheep, which has short, bowed legs. The deliberate production of new breeds and strains, to produce better animals and plants, is commonplace in modern agriculture. It is postulated that such mutations occurred very many times in the last one or two billion years, thus accounting for the evolution of all plants and animals from single-celled organisms.

Occasionally, the term "evolution" is intended to mean no more than the appearance of the various breeds of sheep or dogs and the different strains of flowers. Such evolution is sometimes termed "microevolution," contrasted with "macroevolution," the evolution of an organism from a completely different organism. It is only by means of macroevolution that all plants and animals could have evolved from single-celled organisms. Evolutionists often maintain that the existence of microevolution implies that macroevolution could have occurred. The question is whether or not the large number of mutations needed for macroevolution is possible. The observation of one mutation in a species in a hundred years is not proof that after one hundred thousand years there will have occurred one thousand mutations, with the production of a new, unrecognizable species. The proponents of evolution claim that there has been just such a large number of mutations, whereas those who deny macroevolution maintain that such a large number of mutations, producing a new, unrecognizable species, has not occurred.

The scientific argument against macroevolution can conceivably take two forms. There could be the argument that macroevolution is impossible, regardless of the amount of time allowed. For a given species, it would be maintained that there are limits beyond which change cannot occur. On the other hand, there is the argument that such change is possible in principle, but that there has been far too little time available, even though life may have existed for over a billion years.

Therefore, two questions arise when evolution (i.e., macroevolution, which is always meant in this discussion when the prefix is not used) is debated: Is hereditary change limited or unlimited? From what is known of rates of change, is it possible to show either that evolution could have occurred, or that evolution could not have occurred? These questions appear in different ways in the present discussion.

In defining biological evolution, there is one more problem to be considered. If evolution occurred, *why* did it occur? Why have the simplest forms of life given rise to such a vast number of manifestations of life? The answer usually given is that there has been natural selection and survival of the fittest. When a mutation appeared, it persisted only if the mutation gave the new organism a better chance of survival. What is important is the relation of the mutation to the environment – the climate, the kind of food available, the other living things in the same vicinity, etc.

The environment is not necessarily the cause of the mutation in the organism as a new breed or strain appears. In one sense, the mutations are uncaused, as they are random. Since the original organism must be fairly well-adjusted to its environment, the number of useful mutations is small. One would therefore expect most mutations to make life more difficult; and this is what is actually observed. "Survival of the fittest" is not intended to mean that the process of survival makes the organism more fit.

The environment can occasionally play a role in the production of mutations. For example, cosmic radiation (originating in outer space and found at the surface of the earth) may cause mutation. Such mutation is also random, and the struggle for survival is still not the *cause* of the mutation.

Biblical Evidence

It is assumed here that the Bible considers both plants and animals to possess life, even though the term used in the creation account designating a living creature (*nephesh,* that which breathes) pertains only to animals. This assumption is made because the Bible states that plants die. (Cf. Joel 1:11-12 and Jude 12.)

Plants, created on the third day, were created before animals:

> And God said, Let the earth bring forth grass, the herb yielding seed, and the fruit tree yielding fruit after his kind, whose seed is in itself, upon the earth: and it was so. (Gen. 1:11)

"After his (their) kind" appears nine more times in Genesis 1, with the phrase referring to grass, herbs, trees, marine life, birds, cattle, creeping things, and the other beasts of the earth.

The other facts in Genesis 1 concerning creation are facts the reader possesses, even though he might deduce some of them from scientific study. Why does "after his kind" appear so many times in Genesis 1? The frequent repetition seems to indicate that God is communicating something of special importance. One idea which the phrase "after his kind" communicates to the reader is that what he knows, for example, about cattle *now* begetting cattle, etc., *has also been true from the beginning.*

A claim evolutionists correctly make is that the Biblical "kind" cannot be equated with the modern definition of "species." Since the Bible is for our illumination rather than for our confusion, it is important to know what "kind" means. Perhaps, given only our present knowledge, "kind" cannot be defined accurately. What is so often lost sight of is that "kind" refers to *some* group smaller than the whole group of living things. Therefore, Genesis 1 teaches, for example, that "creeping things" did not by any series of evolutionary changes produce "cattle."

Although there is difficulty today in arriving at the proper definition of "kind," there was some specific meaning for that term (in the Hebrew *min*) as it was used in the Bible. For example, in Genesis 6 and 7 Noah was repeatedly instructed to put into the ark beasts "after their kind" (alternately, "of every sort"). It is sometimes suggested that "kind" is not precise; yet, Noah was continually faced with the question of including or not including an animal in the ark. "Kind" could not have only an approximate meaning for Noah. In addition, in the Levitical law "after his kind" (using *min*) had an unquestionably definite meaning as it referred to animals which the Israelites were not permitted to eat:

> And these are they which ye shall have in abomination among the fowls . . . the kite after his kind; every raven after his kind . . . the hawk after his kind . . . the heron after her kind. (Lev. 11:13-16, 19)

"After his kind" appears in the same chapter several more times, in connection with both forbidden and permitted animals. It seems to be difficult to maintain that "after his kind" has an imprecise meaning.

If we knew precisely what the Biblical "kind" means, then it could be known just what are the boundaries beyond which begetting could not pass. Defining such a boundary would imply not only that there has been no evolution of plants and animals from the simplest forms of life, but it would also provide the limits beyond which change cannot occur. Unfortunately, present fuzziness in defining Biblical "kind" has been used to blur the very real distinction between the kinds. One of the two fundamental questions posed earlier is thus answered. The Bible indicates that unlimited change, from "kind" to "kind," is not possible. In a discussion of what the Bible teaches, the question as to whether or not there has been enough time for the postulated biological evolution, the other fundamental question posed earlier, does not arise.

Many people do not agree with the interpretation of the Biblical "kind" presented here. Bible-believing anti-evolutionists ought to be charitable and honest in this matter. Some of the five evolutionary propositions are clearly settled by the Bible for those who accept its inerrancy. Some Biblical passages the Christian may interpret in only one way; any other interpretation so perverts the Gospel that it is destroyed. Perhaps these "after his kind" passages should not be considered in this category, although it is not to be inferred that even here there is more than one correct interpretation. What *is* wrong, is the idea that the Bible could err on anything concerning biological origins which it teaches.

Further, it is wrong to deduce that since "we know biological evolution is true" (were one to accept such an idea), that the Genesis passages quoted above *must* be poetic or symbolic. No words of the Bible are to be taken as symbolic, or as non-symbolic, just because extra-Biblical information seems to demand it. Whether or not a passage is poetic or symbolic is to be deduced from the Bible itself; and this is, of course, the procedure which is correctly used over and over in Biblical interpretation. The passages cited in which "after its kind" appears cannot be shown, using only the Bible, to be poetic or symbolic. In fact, it is specifically stated in Genesis 2:4 (discussed in the next chapter) that the part of Genesis which includes chapters six and seven, referred to above, is historical.

Scientific Evidence

There has been a tremendous effort to prove, with a lesser effort to disprove, the idea that there has been evolution of plants and animals from the simplest forms of life. Most research biologists have been convinced that biological evolution is a fact. Consequently, the body of biological knowledge which has been accumulated has been fitted to an evolutionary framework. Because new data are usually fitted to the evolutionary framework, those who are not biologists often receive the impression that belief in biological evolution is compelled by the scientific evidence. The existence of this bias does not in itself prove that the evolutionary approach is wrong. On the other hand, fitting data to an evolutionary framework does not prove that evolution actually occurred.

The very intensity of the effort to prove the truth of biological evolution can lead to a situation in which biologists generally reject evolutionary theory. Every conceivable scientific weapon – men, time, and money – has been available to the evolutionist. The powerful faculties at universities in many countries have assumed the truth of evolution. Yet, after a century of effort, a significant minority is uncomfortable with evolutionary theory. This discomfort is worth examining.

A. *Scientific Doubts.* Many biologists, even though they accept evolution, are still not satisfied with the theory. These doubts of a minority do not prove evolution to be incorrect, but they do prove that there are enough questions about evolution so that one cannot maintain that belief in evolution is a scientific necessity. Several examples follow. (Some of these are mentioned by Wiebe.[1])

(1) According to C. Martin of McGill University, the mutation-selection theory (the theory which states that the changes in a species survive only if the changed organisms are better adapted to their environment), the theory almost universally used to explain *how* evolution occurred, seems to be inadequate. Martin is disturbed by

> the almost total lack of scientific caution and self-criticism current in genetical circles, in regard to the accepted theory of evolution by mutation.[2]

One objection made by Martin and others is that mutations leading to a stronger species are not frequent enough.

(2) In a similar manner, many evolutionists are concerned about the detailed mechanism of evolution. They are satisfied with the idea that mutations appeared and that certain of these survived. Nevertheless, J. Rosin listed some problems:

> . . . Life on earth originated, at the earliest, 2 billion years ago. The minimum time required to evolve a new species is about 1 million years. Hence, a mere 2000 species separate the first autocatalytic nucleoprotein molecule from man, a seemingly preposterous conclusion. . . . A great number of species stopped in their evolution, the oldest of them (brachiopod genus Lingula) remaining almost without change for 500 million years. Why?[3]

In answer to the several problems he raised, Rosin proposed a new (a chemical) explanation of how evolution occurred. Others are proposing similar explanations. Regardless of the value of these suggestions, it is noteworthy that after all these years and all the dogmatic assertions concerning the correctness of evolution, evolutionists are still trying to decide how it all came about.

From these quotations, one might suppose that only the means whereby evolution occurred is in question, and that evolution itself is not in doubt. Yet, the very question the evolutionist seeks to answer is, "*How* have living things come to be what they are?" Frequently it has been suggested that the Bible tells us merely *that* God created, not *how* God created. Evolution *by definition* must include the "how." Evolutionists and anti-evolutionists do not disagree, for example, on the structural similarities found so often in the animal and plant kingdoms. The question is whether or not there are lines of descent which account for these similarities. It is not enough to maintain without evidence that one animal descended from another. Such a claim is only an opinion if the mechanism of this descent is not postulated. Furthermore, if the various mechanisms which have been suggested are eventually shown to be very unlikely or impossible, the original claim that descent is a fact becomes weaker and weaker. Without the "how" of evolution, the idea of evolution is but an unsupported opinion.

(3) G. A. Kerkut, of the University of Southampton in England, accepts evolution, but he has shown that the evidence for the traditional picture is weak, and there might have occurred an entirely different kind of evolution. In the preface to his book Kerkut stated,

> May I here humbly state as part of my biological *credo* that I believe that the theory of Evolution as presented by orthodox evolutionists is in many ways a satisfying explanation of some of the evidence. At the same time I think that the attempt to explain all living forms in terms of an evolution *from a unique source*, though a brave and valid attempt, is one that is premature and not satisfactorily supported by present-day evidence. It may in fact be shown ultimately to be the correct explanation, but the supporting evidence remains to be discovered. We can, if we like, believe that such an evolutionary system has taken place, but I for one do not think that "it has been proven beyond all reasonable doubt." In the pages of the book that follow I shall present evidence for the point of view that there are many discrete groups of animals and that we do not know how they have evolved nor how they are interrelated. It is possible that they might have evolved quite independently from discrete and separate sources.[4] (Italics in the original.)

The significance of Kerkut's work for the present discussion lies in his suggestion of a radically different evolutionary alternative. If the chemical and structural differences between some animals are too great to permit the customary descent hypothesis, then the entire evolutionary structure is in difficulty. The traditional evolutionist cannot tolerate Kerkut's suggestion that evolution from *separate* sources occurred, since allowing for such a possibility would be equivalent to admitting that evolutionary theory is not well developed. If one has maintained that species *x* descended from species *y*, and if he has used similarities between these two species to prove that there was descent, it will be difficult for him to change his position and to admit that *x* and *y* are unrelated, having evolved from separate sources. For then the whole procedure used to prove evolution, depending as it does upon structural similarity, would be suspect. Kerkut's alternative, even though it is an evolutionary alternative, cannot be very attractive to the convinced evolutionist.

Kerkut was concerned about the procedures used by the convinced evolutionist:

> It is very depressing to find that many subjects are becoming encased in scientific dogmatism. The basic information is frequently overlooked or ignored and opinions become repeated so often and so loudly that they take on the tone of Laws.[5]
>
> It seems at times as if many of our modern writers on evolution have had their views by some sort of revelation. . . . It is premature, not to say arrogant, on our part if we make any dogmatic assertion as to the mode of evolution of the major branches of the animal kingdom.[6]

(4) J. T. Bonner, Princeton University biologist, said the following in a review of Kerkut's book:

> This is a book with a disturbing message; it points to some unseemly cracks in the foundations. One is disturbed because what is said gives us the uneasy feeling that we knew it for a long time deep down but were never willing to admit this even to ourselves. It is another one of those cold and uncompromising situations where the naked truth and human nature travel in different directions.
>
> The particular truth is simply that we have no reliable evidence as to the evolutionary sequence of invertebrate phyla. . . . If one were to tally the views of experts . . . then one can find qualified, professional arguments for any group being the descendant of almost any other.[7]

Concerning the biochemical evidence for affinities between groups, Bonner continued,

> . . . What we have all accepted as the whole truth, turns out with some mild inspection, to be rather far from it. Apparently, if one reads the original papers instead of relying on some superficial remarks in a textbook, the affinities become extremely clouded indeed.[8]

Bonner is clearly in the camp of those evolutionists who maintain that a radical re-evaluation is needed. He does not suggest that evolution will be discarded as a result of such a re-evaluation. It sems logical to ask of Bonner, Kerkut, and others why it is necessary to assume that evolution occurred if the problems raised are so large. Bonner's answers to this question are extremely interesting, in view of one of our main themes in this book. His answer sheds light on a motive for accepting evolution, and it shows that all men, Christian and non-Christian, tend to accept an important part of the Christian view of the relation beween man and the rest of creation. Bonner neatly expresses man's universal desire for ordering. An important reason for accepting "satisfying information" in the mind of the scientist is

> . . . the grasping of relationships. After all, this is the basis of any sweeping generalization, any general theory. It shows how things fit together; it shows the "system of nature."
>
> Classification is the cure for chaos; we cleave to it just as we avoid the chilly mess of chaos. If we can put animals in sensible groups we enjoy the grouping like a happy, tidy housewife. If the groups have a natural, phylogenetic sequence one to another, we not only have order, but the special joy of logical order. The crowning satisfaction, and hence its extraordinary welcome, was Darwin's theory of natural selection for here there was an explanation of how the phylogenetic grouping could have arisen.[9]

The Christian sees that man has been created in harmony with the rest of creation. Man therefore finds order, just as God intended he should. There is a unity in creation. Apparently man can use this faculty for finding order either to glorify God, the consistent Creator, or to attempt to detract from God's works. The natural man attempts to explain the universe without at any time postulating the existence of God. Evolutionary theory is certainly part of the attempt to take God out of his creation. Even when man attempts to find a substitute for God, he finds order in the universe.

Christians have frequently erred in trying to disprove evolution by attempting to show a *lack* of order among living things. (The limits to the order which is actually found is discussed in [B] below.) A lack of order would probably disprove evolution. But if there is order, it is better to recognize that fact, and to interpret it in a Christian way. Since the Bible teaches what it does, a correct Christian interpretation will not be one which postulates evolution of all living things from the simplest life. A Christian interpretation of the order which exists in the world of living things would be refreshing, and precisely what biological theory needs.

(5) Many other evolutionists have recognized the difficulty of proving evolution. A. S. Romer, a Harvard zoologist, said in 1941,

> The oldest ancestors of the vertebrates are unknown and may always remain unknown.[10]

G. G. Simpson of the American Museum of Natural History, widely accepted as a spokesman of the evolutionary position, said in 1955, in a review of N. J. Berrills' *The Origin of Vertebrates,*[11]

> From about 1875 through the 1920's the origin of the vertebrates was one of the active subjects of evolutionary biology. Then discussion died down from lack of fuel. All the available evidence seemed to be in, and all together was insufficient to warrant much more than a verdict of "not proved." Hardly any new evidence is at hand today, and yet there is room for a reconsideration of probabilities with more perspective than was available when discussion was at its height.[12]

In other words, since the 1920's biologists have claimed greater insight as to the origin of the vertebrates, in spite of the lack of new evidence. Simpson then considers Berrills' admittedly speculative thesis concerning the ancestry of the vertebrates, and he concluded,

> As for the ancestry of the chordates, all is left in darkness without even the dream of 60 years ago.[13]

(6) Doubts among evolutionists are not new. In 1930 A. H. Clark of the U.S. National Museum suggested an alternative evolutionary theory, much as Kerkut proposed an alternative theory thirty years later. Clark was also motivated by the contradiction between the evidence and the accepted ideas concerning evolution. He said,

> . . . The animals of the very earliest fauna of which our knowledge is sufficient to enable us to speak with confidence . . . were singularly similar to the animals of the present day. . . . It is much more logical to assume a continuation of the parallel interrelationships further back into the indefinite past, to the times of the first beginnings of life, than it is to assume somewhere in early pre-Cambrian times a change in these relationships and a convergence toward a hypothetical common ancestral type from which all were derived. This last assumption has not the slightest evidence to support it. All of the evidence indicates the truth of the first assumption.[14]

Clark's alternate evolutionary hypothesis is not of particular interest here. What is of interest is that Clark saw that the problems he raised dealt a blow to the whole concept of evolution. He had said earlier, in 1928,

> Thus so far as concerns the major groups of animals, the creationists seem to have the better of the argument. There is not the slightest evidence that any of the major groups arose from any other.[15]

(7) C. Zirkle, a botanist at the University of Pennsylvania, indicated there is a serious evolutionary problem in his review of *Evolution after Darwin,* twenty essays by specialists in various evolutionary fields.[16] Even though Zirkle was clearly in favor of the evolutionary concept, he pointed out, in connection with the essay, "Morphology, Paleontology, and Evolution," by E. C. Olson (the editor of the journal, *Evolution*) that there is a group of biologists who are silent and

> . . . who are in disagreement with the current theory but who feel it is futile to combat the generally accepted view.[17]

In summary, many responsible biologists have found fault with some of the *crucial* aspects of evolutionary theory, but they still attempt to keep the concept of evolution.

B. *Evaluation.* There were presented earlier in this chapter two basic questions – whether or not hereditary change is limited and whether known rates of change are large enough to allow for evolution. These questions have been answered affirmatively by all who accept the evolutionary idea. Although those quoted above have doubts concerning some aspects of evolution, they have decided that hereditary change is extensive enough to allow for some kind of macroevolution, and that known rates of change are large enough to allow for evolution. These critics seem to say, "There *must* be no limit to the amount of hereditary change which can occur, and there *must* be large enough rates of change to account for the wide variety of living things which exist; but we cannot prove that these things are so." It is *not* enough to know that evolutionists find similarities between organisms; nor is it enough to know that they conclude that simpler structures existed before the more complex. The heart of the evolutionary idea is that there has been *change* from one organism to another. The anti-evolutionist could possibly agree (depending upon the strength of the evidence) that structures are similar and that the simpler is the older. To accept these ideas is not to accept evolution.

One would like to find that biologists are basically agreed concerning the matter of similarity among organisms and the matter of whether or not the older is simpler. If they were agreed, and if evolution were still a scientific option, then the *basic* questions, concerning the limits and rates of change, that is, the questions concerning evolution itself, could be discussed.

Currently among biologists discussing evolution, there is serious question concerning similarities and whether or not the older is simpler. Kerkut (cf. fn. 4) is of the opinion that the great differences among organisms suggests that they evolved from *separate* sources. Others conclude that among the vertebrates, for example, we can never know which evolved from which. It is also stated that "superficial remarks in a textbook" (Bonner; cf. fn. 8) have misled people in this matter of similarities. The "accepted" theory explaining unlimited change, the mutation-selection theory, is not satisfying to some biologists and biochemists whose scientific work has gained for them respect in the scientific world. It is not at all certain, according to some scientists, that there has been sufficient time for the postulated mechanism of evolution to have occurred. In other words, there is doubt concerning the necessary prerequisite for the acceptance of evolution, similarity of structure with the simpler structure the older; and there is doubt concerning evolution itself, which is a process and not a description of living things.

It is not possible in this book, intended for both scientists and non-scientists, to analyze critically scientific data in the way analysis is called for in technical journals and books. What has, however, been attempted here is a presentation of the conclusions of some first-rate experts who *have* carried out this critical analysis. The admittedly minority opinion of these experts carries enough weight to teach caution even to one who does not believe the Bible.

Summarizing the argument thus far, it seems that one who accepts Biblical inerrancy will have difficulty in disproving the arguments concerning the Biblical information which has been presented. If, in addition, one considers the testimony of the evolutionists who have been quoted, he would be dogmatic were he to hold that biological evolution has been *proved*. If the expert scientific testimony were used in court to determine whether or not biological evolution is true, and if, like guilt, truth were established only if it could be demonstrated "beyond the shadow of a doubt," the court would be forced to conclude that biological evolution is "Not proved."

An Alternate View of the Question

It is instructive to attempt to think the thoughts of those who have different opinions, and to see where this leads. Most of those who have studied the question of evolution have tended either to accept or to reject evolution, without adequately considering the other side of the question.

Suppose that the Biblical passages cited earlier did not exist, and suppose further that the Bible contained passages which clearly and in detail taught biological evolution. In those circumstances, Bible-believers would always have known that complex animals were derived from simple ones. Children would have been required to memorize the ten commandments, the names of the kings of Judah, the disciples of Jesus — and also the order of the plants and animals, including which were derived from which. Certainly God *could* have created in this way. The quarrel the Christian anti-evolutionists have with the evolutionists is that evolution is contrary to the Bible, not that God could not have created by means of an evolutionary mechanism. (With respect to the origin of man, not considered in this chapter, the situation is different.)

Thus, if the Bible taught evolution, belief in evolution would have been widespread, and probably not meaningfully opposed, in the year 1800. Biblical knowledge was then an integral part of the culture of most Western countries, even though the number of people who possessed a serious, committed belief in the Christian faith was probably no higher then than it has been at any other time. Biological evolution would have been uncritically accepted if it had been a part of the Biblical revelation.

However, there have been many scientific discoveries pertinent to the study of evolution since 1800. From the beginning of time man has known of the appearance of small, permanent variations – mutations – in plants and animals, and no doubt in this hypothetical situation the Bible-believer of 1800 would have thought mutation to be the evolutionary mechanism. Since 1800 it has been learned that almost all mutations are deleterious. When Martin said there is an "almost total lack of scientific caution and self criticism . . . in regard to the accepted theory of evolution by mutation" (cf. fn. 2), he would have been talking about *Christians.*

The evidence since 1800 has mounted. Evolutionists estimate that only 2000 species separate a single molecule from man (cf. fn. 3). There are "unseemly cracks in the foundations" of evolutionary theory (cf. fn. 7). Among invertebrates, one can "prove" that any group descended from any other, and scientific evidence can never tell us which is the oldest (cf. fn. 7). Animal fossils which have been discovered are the same back to the earliest times, with no indication of change (cf. fn. 14). Worst of all, there is a minority of biologists (in our hypothetical example, a minority of Christians) who know all this and ". . . who feel it is futile to combat the generally accepted view" (cf. fn. 17).

In other words, the scientific results of the nineteenth and twentieth centuries would have upset the complacent applecart of 1800. It would be naive to believe – in the face of all this scientific opposition – that many non-Christians would accept the hypothesized Biblical evolutionary ideas held in 1800. The Christian and the natural man each knows in his heart that if evolution were taught by the Bible, evolution would always be a stumbling-block for those who were asked to accept the Bible. The scientific objections to part of evolutionary theory which have been cited are not well-known. Yet, if these objections were objections to the *Bible,* without doubt they would be common knowledge. One cannot, of course, guess what the natural man would believe concerning the origin of living things if the Bible taught evolution. But it is certain that the natural man would explain repeatedly that the science of the nineteenth and twentieth centuries has shown Biblical ideas to be ridiculous. One would never hear that the Bible correctly anticipated the work of Darwin and his successors. It would, however, be said that what we *now* know of mutations, what we *now* are able to calculate concerning the probability of a molecule being the ultimate "father" of a higher animal, and what we *now* know of the structure of animals "back into the indefinite past" refute evolutionary theory.

How would the Bible-believers respond in such a hypothetical situation, a situation in which they are called upon to defend biological evolution taught by the Bible in the face of some contrary scientific evidence?

Hopefully, Christians would be careful about citing other, seemingly pro-evolutionary scientific evidence. Their case would properly rest on the Bible. Science should be used in only a secondary way when Biblical matters are discussed. Perhaps the more careful Bible-believers would say: "The Bible teaches that God made animals and plants by making the complex from the simple, so that a single cell was the antecedent of all living things. You anti-evolutionists claim that this cannot be so because there is no known way for the necessarily large changes to be accomplished. You can discover no mechanism and, even with some suggested mechanisms, you claim there has not been enough time. You can also claim that there are large gaps in the record of living things, and that we don't find fossils of all the postulated intermediate forms of life which we ought to find if all living things were derived from a single source.

"The Christian, who believes the Bible, has an answer for you. Merely because *man* cannot find a mechanism for biological evolution and because *man* cannot prove the intermediate forms of life ever existed, it is not thereby proved that evolution did not occur. God said he made things by an evolutionary process, but he did not say man had to be able to understand. What man cannot fathom or discover means nothing. Man cannot necessarily understand God's ways. God performed many miracles, and he tells us about them in the Bible. Man ought not to presume that he will be able to understand all these miracles. Perhaps man will never be able to understand biological evolution. We might just as well admit that creation by evolution is a miracle. The Christian understands that the universe is not self-contained, that it was created by God, and that it is sustained by him. Many of his sustaining acts which man has observed do not fit into man's natural law; those acts are miracles. Christians agree that it seems scientifically impossible that monkeys and elephants were derived in a few billion years from the simplest forms of life. It also seems impossible that Christ's words could bring Lazarus from the grave. Nor do we understand how manna could fall from the skies. Yet, we believe these things occurred because God tells us they occurred. God has the power to do what he says he did. All your we-don't-see-how-it-is-possible arguments against evolution cannot possibly affect the Christian. What *man* can understand is quite irrelevant."

What does this reversal of the usual roles accomplish in this discussion? This reversal shows that the scientific arguments against evolution – deleterious mutation, insufficient time, lack of evidence that the intermediate forms existed, lack of a convincing mechanism, etc. – add up to a requirement of a large amount of faith on the part of whoever accepts evolution. This reversal shows that the kind of faith needed to accept evolution is very similar to the kind of faith the Christian has as he accepts miracles. The scientific evidence adduced for evolution, such as the similarity of certain structures, man's inability to assign functions to certain "vestigial" organs, and the existence of mutations, is only circumstantial evidence. It is evidence which can be fitted to an evolutionary theory, but it is also evidence which can be fitted to another theory – special creation. Furthermore, this evidence does not provide the answer to the *one* question that a comprehensive theory of evolution must eventually provide, the answer to the question concerning mechanism.

Doubtless many scientists who accept biological evolution are quite willing to admit the existence of the problems which have been mentioned, and they will admit that faith is indeed required. Is it undesirable to require faith?

In the last chapter the question of faith was raised in connection with the discussion of the production of the simplest life from non-living matter. It was concluded that the scientists who describe the evolution of the simplest life from non-living matter actually describe a miracle. Their "spontaneous" evolution of the simplest life from non-living matter is just as "spontaneous" as the transformation of water into wine at the wedding in Cana of Galilee. *In the same way, modern science has revealed enough about the history of life and the mechanism of reproduction to indicate that any biological evolution from the simplest life would also have to be termed a miracle.*

If belief in evolution from the simple to the complex forms of life is belief in a miracle, why should such a belief be accepted? Why accept a miracle not taught by the Bible? Contrary to the hypothetical situation described above, the Bible does not instruct us to accept evolution. "After its kind" seems to the Bible-believer to rule out evolution. Even for those who cannot see that point, what positive reason is there to accept evolution? To accept either creation or evolution calls for belief in a miracle. There is then no remaining objection to the idea that at various times God created new kinds, and that these kinds always reproduced individuals like themselves.

In the previous chapter it was shown that in the last analysis only a miracle can account for the origin of the simplest life. It is not too important to know whether or not God used non-living matter to make the simplest life. In the same way, it is possible that God created new kinds by performing miracles on existing kinds. Whether or not existing organisms were used as starting materials is quite unimportant. Creation of the kinds was miraculous, and once they were created, they reproduced naturally after their own kind.

REFERENCES

1. Wiebe, H. T., *Journal of the American Scientific Affiliation, 18*, 112 (1966).
2. Martin, C. P., *American Scientist, 43*, 100 (January, 1953); quoted by Wiebe
3. Rosin, J., *Chemical and Engineering News* (Letters), March 27, 1967, p. 6.
4. Kerkut, G. A., *Implications of Evolution*, Pergamon Press, Oxford, 1960; pp vii-viii.
5. *Ibid.*, p. viii.
6. *Ibid.*, p. 155.
7. Bonner, J. T., *American Scientist, 49*, 240 (June, 1961).
8. *Ibid.*
9. *Ibid.*
10. Romer, A. S., *Man and the Vertebrates*, Univ. of Chicago Press, Chicago, 194 quoted by Wiebe.[1]
11. Berrills, N. J., *The Origin of Vertebrates*, Oxford Univ. Press, Oxford, 1955.
12. Simpson, G. G., *Science, 122*, 1144 (1955).
13. *Ibid.*
14. Clark, A. H., *The New Evolution Zoogenesis*, Williams and Wilkins, 193 quoted by Wiebe.[1]
15. Clark, A. H., *Quarterly Review of Biology*, 52 (Dec. 1928); quoted by Wiebe
16. *Evolution after Darwin* (Sol Tax, ed.), *Evolution of Life*, vol. 1; Univ. c Chicago Press, Chicago, 1960.
17. Zirkle, C., *Science, 131*, 1519 (1960).

13 *The Bible on the Origin of Man*

Many who deny the fifth proposition, "Man's body evolved from animals," maintain that seldom has Satan devised a more wicked idea. For them, the concept of man, the relation of man to God, and the work of Jesus Christ, the Incarnate Man, are all at issue in this question. The question of man's origin is doubtless the most controversial part of the evolutionary idea. A discussion of *evolutionism,* the world- and life-view which depends upon the acceptance of all five evolutionary propositions, is most intimately associated with the fifth proposition. Evolutionism is concerned with the nature of man, of his religion, and of God. In this book, only a few aspects of evolutionism are treated and even this treatment is postponed until after the discussion of the five propositions. Unfortunately, the debate concerning evolution is often badly confused because the debaters consider simultaneously the five propositions, or their equivalent, and the philosophical implications of those propositions. Discussion is more relevant if discussion of the philosophical implications is deferred until after a decision is made concerning the five basic propositions.

It has been maintained that man did indeed evolve from animals, and that evidence for such an evolution is that no other origin of man so far proposed is conceivable. Accepting human evolution for this reason seems to be accepting it because certain consequences of the evolutionary idea are desired. For example, if man evolved, it is unnecessary to accept instantaneous creation of man, the alternative which is held to be inconceivable. Instantaneous creation cannot possibly have a place in the thinking of the natural man. Another consequence of accepting human evolution is the idea that man is constantly improving. There is the bright hope of unending progress. For many, the ideal for man is a continual upward climb, with restraint neither from the past nor from any force outside the observable universe. Ancient laws formulated to regulate human behavior are therefore no longer relevant. Irrespective of the desirability of these consequences, they are entirely irrelevant if the fifth proposition is untrue.

Therefore, it is appropriate to postpone a discussion of the consequences of either the evolutionist or anti-evolutionist point of view until the question is decided, on a basis other than desirability.

The Concept of Animal-to-Man Evolution

The question must be defined so that the views of both the "theistic" and the "atheistic" evolutionist are discussed. (These labels are used only for convenience; the questions concerning whether "theistic" and "evolutionist" are contradictory, and the relation between the positions of the theistic evolutionist and the natural man are not discussed at this time.) If the position of the theistic evolutionist is carefully discussed, the essential points of the position of the atheistic evolutionist will thereby be automatically considered. To state this, however, is not to accuse the theistic evolutionist of atheism.

The theistic evolutionist emphasizes that man's body, not his soul, evolved from animals. The anti-evolutionist insists that man was created in the image of God. The theistic evolutionist states – correctly– that "image" refers to knowledge, righteousness, and holiness, all spiritual attributes of man. In limiting the discussion to the body there is no intention to promote an un-Biblical dualism of soul and body. One is not on firm debating ground when the weight of the argument is put on the mysterious unity of soul and body. The discussion is thus limited to the part of man which is of interest to both theistic and atheistic evolutionists – the body.

Biblical Evidence Concerning the Origin of Man

Is the teaching of the Bible consistent with the idea that man descended from animals? It is generally agreed that the key passage is

> And the Lord God formed man of the dust of the ground, and breathed into his nostrils the breath of life; and man became a living soul. (Gen. 2:7)

The theistic evolutionist maintains that the word "dust" in this verse refers to an animal, and that God put a soul into an animal body to make man. Obviously, if it could be decided just what "dust" means here, Bible-believers would have the desired answer. It is doubtful that a convincing answer to the main question posed by the concept of theistic evolution can be obtained in this way. The Hebrew word here used for "dust" appears over one hundred times in the Old Testament. The frequency of the appearance of this word, and the difficulty of determining what it means in some of its uses, suggest that other methods of attacking the question may be of more help.

The Biblical evidence concerning the origin of man is presented here in a discussion of three statements. An attempt is made to show that these three statements are Biblical truths, and that taken together they decide the question of theistic evolution. The first statement concerns the nature of the body of man, and the second and third statements pertain to the key passage already quoted, Genesis 2:7.

A. *Nothing Else in Creation Is Like the Body of Man.* If a Bible-believer is convinced that this statement is wrong and that the bodies of men are similar to the bodies of animals, he might attempt to prove his point from the Bible, by citing,

> I said in mine heart concerning the estate of the sons of men, that God might manifest them, and that they might see that they themselves are beasts. (Eccl. 3:18)

Is it possible that men are beasts? The Preacher, who wrote Ecclesiastes, continues:

> For that which befalleth the sons of men befalleth beasts; even one thing befalleth them: as the one dieth, so dieth the other; yea, they have all one breath; so that a man hath no preeminence above a beast: for all is vanity. All go unto one place; all are of the dust, and all turn to dust again. (Eccl. 3:19-20)

The Preacher then wrote concerning life after death:

> Who knoweth the spirit of man that goeth upward, and the spirit of the beast that goeth downward to the earth? (Eccl. 3:21)

Up to this point, he who maintains that man's body is not unique, that it is like that of a beast, seems to have the argument all his way. Man's body, like the body of the beast, is in the dust, says the Preacher. But at the end of his book the Preacher said something which might motivate the reader to investigate further:

> Then shall the dust return to the earth as it was: and the spirit shall return unto God who gave it. (Eccl. 12:7)

Not the whole of God's plan is given in Ecclesiastes. Yet, there is hope in the verse just quoted and also at the end of the book:

> Let us hear the conclusion of the whole matter: Fear God, and keep his commandments: for this is the whole duty of man. For God shall bring every work into judgment, with every secret thing, whether it be good, or whether it be evil. (Eccl. 12:13-14)

The Preacher ended his message. But he raised some questions which he did not answer. Incompleteness in one book of the Bible is not unusual, since the Bible — even though it was written by many men — was inspired by one Holy Spirit. The Preacher left the bodies of beasts and men in the dust, but the *spirit* of man went upward. The Preacher does not deny the statement, "Nothing else in creation is like the body of man." Nor does he confirm it. Evidently he had another purpose.

When the Ecclesiastes passages are taken together with other parts of the Bible, it is seen that man's body is indeed unique. The Holy Spirit who inspired the Preacher to write that the body of man returned to the dust also inspired Daniel to write of dust:

> And many of them that sleep in the dust of the earth shall awake, some to everlasting life, and some to shame and everlasting contempt. (Dan. 12:2)

Thus the Bible leaves the beast in the dust, but it teaches that man will rise from the dust. Man will rise not because God will make him perfect, since after all some will rise to shame and everlasting contempt, but rather because his body is the body of *man*. The combined teaching of the Preacher and Daniel indicates that man's body and the body of an animal are different. It is easy to observe similarities between the anatomy of man and the anatomies of certain animals. The questions posed by the concept of theistic evolution are not answered by comparing anatomies. It is quite relevant, when considering the nature of the body of man, to include whatever is known of the *destiny* of his body. The Biblical teaching that the body of each man will exist forever provides one reason for the conclusion that nothing else in creation is like the body of man.

The uniqueness of man's body is also taught in the New Testament:

> All flesh is not the same flesh: but there is one kind of flesh of men, another flesh of beasts, another of fishes, and another of birds. (I Cor. 15:39)

The analogy of the different kinds of flesh is one of the analogies Paul used in I Corinthians 15 as he discussed the process whereby the earthly body, which goes to the dust, becomes the risen body. He wrote that the various kinds of flesh are as different from each other as the earthly body of man is different from his risen body.

In the other analogies Paul compared the difference between the earthly body and the risen body to the differences between the seed and the plant, and the terrestrial body and the celestial bodies. One might maintain that there are not *absolute* differences in Paul's analogies. Thus, there is a similarity as well as a difference between the earth and the celestial bodies. A complete discussion of Paul's rather complex argument would be long and involved, and perhaps in such a discussion it would be decided that the use of I Corinthians 15:39 is not absolutely conclusive for the point being made.

Yet, the *least* that can be said concerning this passage is that it casts considerable doubt on the idea that there is only a qualitative difference, as contrasted with an absolute difference, between the body of man and the bodies of various animals. Are not, asked Paul, men, beasts, fishes, and birds different? Just as different is the risen body of man from his earthly body. The risen body of man is a continuation of his earthly body, but the risen body *will live forever*. So in another sense the risen body will be strikingly different — different in an ultimate, absolute sense — from the earthly body. Paul proves this *absolute* difference by referring the reader to the differences between the bodies of beasts, birds, fishes, and men. Here too there is similarity, but Paul assumes his reader knows of the absolute differences. If his reader could not be assumed to know of these absolute differences, the analogy would fail.

This passage seems to put the burden of proof on those who deny absolute differences between the bodies of men and various kinds of animals. Is man a higher mammal, related to certain higher mammals? How can he be related to higher mammals, if the difference between the body of man and the bodies of beasts is as vast as Paul assumes the reader knows it to be?

It has been emphasized in this book that the science the Bible assumes to be true must be true science. The I Corinthians 15:39 passage seems to teach that there is an absolute difference between the bodies of men and animals. Proceeding with *this* assumption, the Bible teaches us certain facts about our future life. If the risen body of man is to be as similar to our earthly body as the theistic evolutionist says our earthly body is similar to that of certain animals, where is the comfort of the Christian? Could the Christian possibly possess inner peace and comfort concerning the future if he thought his body would improve only as much as his earthly body is supposed to have improved over that of the highest animal?

B. *God Gave Man Life After He Gave Animals Life.* The second statement in the discussion of theistic evolution is derived from Genesis 2:7, the central statement concerning the creation of man, quoted above. What is meant by *nephesh,* translated in this passage as "living soul"? The theistic evolutionist maintains that God added a soul to a previously existing animal body. Does Genesis 2:7 describe this soul-adding process?

Nephesh is a general term, meaning "living creature," and is used as such in the following passages:

> And God said, Let the earth bring forth the living creature after his kind, cattle, and creeping thing, and beast of the earth after his kind: and it was so. (Gen. 1:24)
>
> And the bow shall be in the cloud; and I will look upon it, that I may remember the everlasting covenant between God and every living creature of all flesh that is upon the earth. (Gen. 9:16)

Thus, *nephesh* can refer to animals or to men and animals together. Following is Genesis 2:7 after *nephesh* is translated "living creature":

> And the Lord God formed man of the dust of the ground, and breathed into his nostrils the breath of life; and man became a living creature. (Gen. 2:7, last word different from Authorized Version)

When the Bible states here that God breathed into man's nostrils the breath of life, it teaches that man received life at this time. In the same verse the words "man became a living creature" indicate that at this time man received life. Thus, in one verse the reader is told *twice* that man received life at this time.

It does not seem possible that man could have descended from animals, with the creation of man consisting of the addition of a soul to an already existing animal. The passage cited does *not* mention the soul of man, but it does state twice that man then received life. The life man received is not here described as different from the life of an animal; otherwise *nephesh,* used also to describe the life of animals in the passages quoted above, would not have been used. It seems that the theistic evolutionist loses the principal reason for his view when "soul" is removed from Genesis 2:7.

Furthermore, it is precisely the same Hebrew word, *nephesh,* "living creature," which gives *positive* support to the idea that man did not descend from animals. The evidence here does not indicate that God took an animal and added a soul to make man, but there *is* evidence in this verse indicating that man was made from non-living matter. It is stated twice in this verse that God gave man life *after* animals were given life. A descendant of an animal could not be given life; such a descendant would already possess life. (Another possibility is that God added life to a dead animal to make man. No one seems to accept this view, and it is probably an unimportant view.)

It is interesting to speculate (although such speculations must be made with great care) what the history of the evolution controversy within the Christian church would have been had the English translators (who could not have foreseen the controversy over theistic evolution in 1611) not used "living soul" in Genesis 2:7. No translation would, of course, be very important for those who believe the passage is symbolic. But it seems very likely that had "living creature" been used, many more defenders of the uniqueness of the body of man would have understood the true meaning of the verse. At the same time, theistic evolutionists would have been denied Biblical support of the one idea they *must* derive from the Bible, namely, the idea that the creation of man is to be identified with the creation of his soul, not his body. In what other way could the theistic evolutionist have derived from the Bible the admittedly theological idea that the soul and body of man were created at different times?

There is still another difficulty with the sequence of events proposed by the theistic evolutionist. There is adequate Biblical proof that death has come to mankind because of sin. For example, Paul stated,

> Wherefore, as by one man sin entered into the world, and death by sin; and so death passed upon all men, for that all have sinned. (Rom. 5:12)

It is certain that "death" here includes the *physical* death of man. It is not possible that man faced the prospect of physical death before he sinned. The Bible teaches in an almost endless variety of ways that man's misery, including his physical death, is in the world because of man's sin.

But sin is a breaking of the moral law, and therefore sin cannot be associated with the body of an animal. Sin, to employ the description of man used by the theistic evolutionist, is associated with man's body *and* his soul. If body-soul man existed only after the event described in Genesis 2:7, then sin, the cause of the physical death of man, could have occurred only after that time.

If the theistic evolutionist is correct, how is it possible that man faced the prospect of physical death only after the Genesis 2:7 event? Did certain animals, descended from animals which had known death, become free of death when they were transformed into men? Did these animals (or animal, depending upon how many are supposed to have become men), after having been transformed into men, receive a temporary immortality, only to lose it when they sinned? There is no suggestion in the Bible that there was a temporary immortality. Had man descended from animals, one would expect that man would die just as all his ancestors had died. Before the fall man had not sinned, but neither had his ancestors; they died, and therefore he would also die. Yet, the Bible states over and over that man dies because of something he did. If God indeed had given man temporary immortality, the only way we could know that it was given is, strangely enough, from those passages which tell us that immortality was taken away. If theistic evolution is not accepted, and if it is understood that body-soul man was created in the image of God, one would *expect* that this body-soul man would be immortal. The idea of immortality of the whole man follows if it is not assumed that body and soul were created at different times. Only then is it possible to understand why the Bible tells us of the removal of the immortality of the body, without telling us also that this immortality had previously been given.

Perhaps the temporary immortality idea is such a bizarre idea, that showing it to be the logical outgrowth of theistic evolution will help some to understand that there is no Christian alternative to the idea that the "dust" from which man was made was non-living matter.

C. *The Account of the Creation of Man Is Not Symbolic.* The theistic evolutionist denies that God gave man life after he gave animals life. He claims that the key passage, Genesis 2:7, is symbolic, teaching *that* God created, not *how* he created.

When the "symbolic" claim is made for a passage, it is sometimes difficult to decide whether or not such a claim is justified. In this instance there is no problem in deciding whether or not the "symbolic" claim is justified. The passage which contains Genesis 2:7 is virtually labeled, "This passage is not symbolic." The first thirty-four verses of the Bible describe the seven days of the creation "week." The thirty-fifth verse introduces us to the rest of Genesis and in fact the rest of the Bible:

> These are the generations of the heavens and of the earth when they were created, in the day that the Lord God made the earth and the heavens. (Gen. 2:4)

The word "generations" is associated with history. The key passage concerning the creation of man, Genesis 2:7, is but three verses later. The event of Genesis 2:7 is as much a part of human history as, for example, the death of Jacob recorded in the forty-ninth chapter of Genesis. There is no point between Genesis 2:7 and Genesis 49 at which one can say,

"Here symbol ends and human history begins." Genesis 2:4 rules out that possibility. What precedes Genesis 2:4 is, of course, also historical; but at Genesis 2:4 there begin the events (so different from the grand events of God's creation week), which comprise the story of man, his *generations*, in the Bible. To believe that Genesis 2:7 is symbolic, not historical, is to miss the thrust of the passage which begins with Genesis 2:4.

Since man's body is different from that of animals and God gave man life after he gave animals life, the creation of man was unique.

14 Evolution: Conclusions and Consequences

In previous chapters various aspects of biological evolution were discussed. By examining relevant Bible passages it was found that the Bible teaches that man did not descend from animals. With respect to the scientific aspect of the question, the objection to human evolution is the same as the objection to animal-to-animal and plant-to-plant evolution. In the case of human evolution some scientists have presented a circumstantial case, just as a circumstantial case has been presented for animal and plant evolution. But they have not presented a convincing mechanism whereby the large gap between animal and man was crossed. Accepting human evolution from animals is to accept, in order that the gap between animals and man be crossed, the equivalent of a miracle.

Thus, there is no need to discuss separately the scientific "evidence" for the descent of man from animals. But the Biblical evidence concerning the origins of animals and plants differs from the Biblical evidence concerning the origin of man. The Bible may allow for a miraculous conversion of one animal or plant into another animal or plant, but it does not allow for the miraculous conversion of an animal into man. Taking the Biblical and the scientific evidence together, it is therefore concluded that (1) the only non-human evolution which could have occurred involves miracles and (2) there has been no human evolution. This is the general conclusion made in this book concerning biological evolution.

Theistic Evolution

In discussing the Biblical evidence concerning the origin of man, the position of the theistic evolutionist was found to be in error. Theistic evolution can now be examined more fully. A more complete examination of theistic evolution is presented because the theistic evolutionist rejects the idea that one's position is basically either that of the natural man, or that of the Christian as it has been outlined in this book. The attempt of the theistic evolutionist to find another Christian alternative to the position of the natural man is extremely important because of the opinion which the theistic evolutionist has of the Bible. The theistic evolutionist claims that the Bible cannot provide scientific information. If his attempt had succeeded, the main thesis of this book, namely, that the Bible can provide such information, would be incorrect. We would then, for example, be uncertain concerning the space-time finitude of the universe, and the Christian's world-and-life view would be changed significantly.

It is now attempted to show that the theistic evolutionist has failed precisely because he has sought a Christian alternative which rules out use of the Bible in scientific questions. The procedure used is to consider the answers of the natural man, the theistic evolutionist, and the Christian who allows for "science in the Bible," to the following question:

According (1) to the Bible and (2) to the scientific evidence, did man's body descend from animals?

Answer of the natural man: (1) *Biblical evidence:* the Bible can provide no information because it claims to be of supernatural origin, and the supernatural does not exist. (2) *Scientific evidence:* the evidence that man did descend from animals is scientific. *Nature of the evidence:* man's body is similar to the bodies of some animals. *Evolution mechanism:* gaps in the evolutionary chain were crossed by chance mutations, without supernatural guidance.

Answer of the theistic evolutionist: (1) *Biblical evidence:* The Bible can provide no information because it does not answer scientific questions. (2) *Scientific evidence:* the evidence that man did descend from animals is scientific. *Nature of the evidence:* man's body is similar to the bodies of some animals. *Evolution mechanism:* gaps in the evolutionary chain were crossed by God's guidance, as he performed the equivalent of a miracle.

Answer of the Christian who allows for scientific truth in the Bible: (1) *Biblical evidence:* the Bible teaches that man did not descend from animals. (2) *Scientific evidence:* scientific evidence related to the question does not actually answer the question. *Nature of the evidence:* man's body is similar to the bodies of some animals, but this is not proof of evolution.

An analysis of these three answers, with special attention given to the differences in the answers, reveals that the theistic evolutionist's position is indefensible. The theistic evolutionist and the natural man both rule out Biblical evidence, although for different reasons. They both accept evolution because man's body is similar to the bodies of some animals. The only difference in their two positions lies in the evolution mechanism they accept: the natural man postulates chance mutations, whereas the theistic evolutionist states that God's special, miraculous guidance was necessary, inasmuch as chance was not enough.

The idea that God guided these changes in a special way is an idea which the theistic evolutionist *cannot* derive, *using his own assumptions,* from either the Bible or science. He cannot derive from the Bible the idea of God's guidance across a gap because in his view the Bible cannot teach such a scientific matter. Nor can he derive the idea of God's guidance across a gap from science, as he conceives of science: the crossing of a gap is equivalent to the occurrence of a miracle, and he does not suppose that science can demonstrate that a miracle occurred.

In other words, the theistic evolutionist has permanently prevented himself from proving the truth of the one idea which makes his position unique, namely, the idea that God guided life across gaps in a miraculous manner. Thus, the theistic evolutionist depends upon an unprovable assumption.

Is it undesirable to have faith in an idea not taught by the Bible? There are obviously many such ideas which are accepted and which are permissible. Yet, in this instance an *unprovable* idea is accepted. There are alternatives to accepting this unprovable idea, and the theistic evolutionist cannot logically object to the substitution of *another* unproved idea for his unproved idea. Furthermore, if one could reject his unprovable idea in favor of a proved idea, theistic evolution would crumble. Thus, theistic evolution is forever vulnerable. Its one distinctive idea, using its own premises, can never be more than a guess.

What happens if the miraculous guidance idea of theistic evolution is discarded? (1) If it is discarded although the "circumstantial evidence argument" for evolution is kept, then the positions of the theistic evolutionist and the natural man are indistinguishable as far as biological evolution is concerned. It does not matter which evolutionary mechanism is accepted, providing the element of the miraculous is not accepted. The position of the natural man is that there was no miracle, and no matter what is substituted for miracle by the theistic evolutionist he would then not contradict the position of the natural man.

(2) There is another course the theistic evolutionist can take. He can seek to prove that God guided evolution miraculously. To prove that a miracle occurred, he must be willing to use the Bible, and to drop the assumption that the Bible does not provide scientific information. Then the unprovable assumption becomes provable, although not necessarily proved. The theistic evolutionist would then go to the Bible to determine what it teaches about the origin of the body of man. He would then find that the Bible teaches that man did not descend from animals. He would then have accepted the Christian idea concerning the origin of man which is being defended here.

A Challenge to Christian Scholars

Concerning man, the discussion in this book has focused on the meaning of Genesis 2:7, and on whether or not the biological steps postulated between animals and man constitute the equivalent of a miracle. Other questions about man have dominated the discussion of evolution in recent decades. For example, the question of the age of man has been considered important. What was man like in his early years? Where did he live, and how has he come to inhabit most of the earth? What was his early culture like – i.e., what tools did he use, what did he eat, how did he amuse himself, and what was his religion? Usually these questions are asked with the implication that man has indeed evolved from a more primitive state.

Even though man did not evolve, it does not follow that the questions about early man are of no interest to the Christian. Rather, these questions should be re-phrased with the assumption that man did not descend from animals. Thus, it might be asked, when did God create Adam? What were Adam, Abel, Cain, Seth, Enos, and their contemporaries like? Where was the Garden of Eden? Where did men live before the Noahic flood? How far abroad did men live before the tower of Babel was built? What is the relationship between the culture described in Genesis 4 – where there is reference to the first tent-dwellers, the first users of musical instruments, and the first brass and iron workers – and modern archaeological discoveries concerning these things?

Discussing such questions from either the point of view of the evolutionist or the anti-evolutionist does not help answer the question of the truth or falsity of the idea that man descended from animals. One cannot talk meaningfully, for example, about the religion of early man unless it can be decided which of these is true: man slowly accumulated religious ideas, improving his religion all the while; or, Adam, the first man, received direct revelation from God and began with true religion, with man's worship then degenerating for a long period of time.

The answers to these interesting questions concerning early man can edify the Christian. He cannot answer these questions adequately without use of the Bible. Here also it is seen that it is wrong to assume that the Bible teaches no scientific fact. The archaeologist interested in the first tent-makers and other culture-formers listed in Genesis 4 cannot carry out his work properly if he does not accept the facts given in that passage as absolute truth. Few workers in this field have assumed that man did not evolve, and therefore the correct analysis of the relevant archaeological data remains to be given. Making such an analysis, using the proper assumptions concerning man, is a challenge to Christian scholars.

Consequences of Beliefs Concerning Evolution

For both the evolutionist and the anti-evolutionist the most important questions concerning the history of the universe are questions concerning man. From where has man come? Where is he going? Without such questions, the concept of evolution would not be as important as it is.

It is instructive to observe how men answer questions related to the origin, history, and future of man. Observing how men's philosophies are affected by their acceptance or rejection of evolution, does not enable one to decide upon the truth or falsity of evolution. Since in the previous chapters conclusions concerning the five evolutionary propositions have been made, an observation concerning what evolutionary theory does to modern thinking can now be added.

It was shown earlier that if man's scientific theory does not conform to the real universe, he sooner or later encounters serious difficulty. This is true for the natural man's assumption that creation from nothing is impossible; and it is also true for his assumption that man has evolved from animals. The seemingly impossible situation in which the natural man finds himself when he accepts those other incorrect ideas was discussed. It can now be shown that an impossible situation is developing because of the widespread acceptance of the idea of human evolution.

A. *The Hitler Fallacy.* One of the ideological supports of the Nazi party during the years it ruled Germany was the idea that one (so-called) race of men is superior to all others. Hitler himself especially emphasized this idea.

Men have always held to an idea similar to this one. Yet, the modern evolutionary scientist inadvertently gives his blessing to the idea that one group of men has evolved further than other groups. For, who can deny the super-race idea if man has indeed evolved? How can one assume that man has evolved for a few million years, without also assuming that some men have evolved further than others? The nature of the physical environment is supposed to be important in determining evolutionary rate, since the relation between a mutation and the environment varies according to environment. But the climate and other environmental factors vary considerably over the surface of the earth, and therefore one would expect some divergence among men.

Evolutionary anthropologists admit there is divergence with respect to body type, color, etc., but they almost always deny that there are intelligence deficiencies in some nations, ethnic groups, and races. Why make such a denial? If men evolved from animals, men were certainly once less intelligent and less teachable than they now are. If the evolutionary process brought differences with respect to color and body type, it would be fortuitous indeed if the capacity to learn changed uniformly among all men, once the evolutionary march away from the animal state was begun.

Almost all evolutionary anthropologists vehemently deny this conclusion concerning intelligence, but their denials have not been heard by Hitler and the followers of the Hitlers of this world. A non-scientist is in danger of accepting the Hitler super-race idea if he does not accept the Biblical approach to the study of man. Observe how this comes about. The evolutionist postulates that there are basic differences – except for the difference of "capacity to learn" or the "capacity to progress" – in men because of the evolutionary process. The non-scientist accepts the idea of basic differences not so much because of what the evolutionist has said, but because he observes color and other differences. The non-scientist might not be careful to make the exception of "capacity to learn" or "capacity to progress." In this way, the improper idea of "racial" superiority gains acceptance.

If the non-scientist thus accepts a poor idea which most evolutionists do not teach, is it the fault of evolutionary theory? When Hitler and his followers uttered their untruths about a master race, opposition by scientists who accepted evolution was on weak ground. Evolutionary science could not give an adequate reason for acknowledging the equality of all men. The evolutionist could not teach that for both man and animals change is limited, and that both man and animals reproduce after their kind. The evolutionist could not insist that there has *not* been a long march from "idiot" animals to a spectrum of modern races of man, capped by some super-race. Nor could the evolutionist teach that men did not descend from many ancestors, but that instead men descended from one man, Adam. Therefore, no nation, ethnic group or "race" is inferior, if "inferior" means "capacity to learn" or "capacity to progress." Even though most evolutionary anthropologists cannot be accused of accepting the idea that some men are inferior, they have fallen down badly in their failure effectively to oppose this idea. They have usually assumed equality among men without evidence. Anthropologists themselves admit that the evidence which *has* been offered is not adequate to decide the question of equality. Let the evolutionary anthropologist answer one question: *If all men are equal, why are they equal?*

B. *Racism.* Hatred between races of men, where a "race" is popularly defined by color, is world-wide. Racism focuses not so much on the idea that one race is *superior* to all others, but on the idea that one race is *inferior* to all others.

It is probably expected in this scientific age that some racists would attempt a scientific rationalization of race hatred. Carleton Putnam vigorously attempted to prove that the Negro is inferior to other races, particularly the "white race," in *Race and Reason: a Yankee View.*[1] He quoted Toynbee's remark that there has not been enough time to prove that the Negro has the capacity for civilization, and that 500 billion years will be available for the Negro to prove his ability. Putnam responded,

> On such a time scale I would agree with Toynbee. In the next 500,000,000,000 years I would be quite prepared to concede the possibility the Negro may, through normal processes of mutation and natural selection within his own race, eventually overtake and even surpass the white race.[2]

Apparently the evolutionary process could overcome the Negro's inferiority. After Putnam's book appeared, he was widely hailed in the part of the South in which racism was particularly strong. Many who wanted to believe the Negro is inferior repeated the idea that evolution explained the difference between the races. In the wake of Putnam's book and his personal appearances in the South, there was a public discussion of human evolution, and the idea was advanced that the Negro evolved 200,000 years later than the white man.

The scientific community cannot be held responsible for such wild statements. Yet, the record would look much better if the scientific community – in this case, the community of evolutionary anthropologists – had a record of teaching that which faithfully corresponds to the structure of the universe. God could have created a whole spectrum of beings between animals and man, with no clear line of demarcation, but he did not. If it is assumed he did create in that way, a science and a world-and-life view will be built on the wrong foundation. Men encounter difficulty here even as, for example, physicists encountered difficulty in the years just prior to the advent of quantum mechanics, an event which enabled physicists to understand the physical world far better than ever before. Similarly, there will be racial tension until men completely reject the idea that man descended from animals.

C. *Marxism.* Public opinion of Marxism is often determined by the behavior of the governments of present-day Communist countries. Opposition to Communism as a political force usually exists because of Communistic economics and the violence which attends the spread of Communism.

If that were all there is to Communism, there would be no reason to include Communism in a discussion of science and the Christian faith. After all, history presents us with a very long list of tyrannies in which individuals were reduced to slaves of the state, and in which violence was common. But Communism is different. Even though we might feel that Communist rulers often act more like gangsters than responsible citizens, they are gangsters with a comprehensive Marxist philosophy which motivates them with a holy zeal. One underlying idea makes Marxist philosophy possible. Without this one idea there would be no Marxism, and consequently no Communism with its coherently organized thrust towards world domination. The all-important practical consequence of the one underlying idea of Communism is that the sustained zeal of large numbers of hard-core Communists throughout the world, all working towards the same end, is a force almost unique in world history. These Communists are so sure of their philosophy that they will suffer martyrdom.

What is the mainspring of this philosophy, the underlying idea of Marxism, and consequently its political manifestation, Communism? *The one idea which makes Marxism possible is the idea that man can create a perfect society.* The Nazi and the racist look back to what they believe has happened, and they see man today existing on various levels. The Marxist focuses his attention on the future, when man will have improved his society immeasurably. This Utopian idea has had such an effect on men because of the evolutionary optimism which began at about the same time as Marxism, in the middle of the nineteenth century. In Marxist theory, man will control his own development by political and other means, and the Marxist includes in "other" means the possibility of controlled human mutations. The inclusion of the use of biological evolution is not so important; but what is important is that the entire Utopian idea was given encouragement by the idea of human evolution. Without the idea of the achievability of this perfect state, the Communist's idea of progress and plans for the future collapse. If the Communist abandons the idea of the inevitable march towards perfection, Communism becomes just another tyranny.

The evolutionary anthropologist usually denies that Marxism is true. Yet, the concept of limitless human change espoused by the evolutionary anthropologist has provided an ideal starting point for Marxism. If scientists for the last century had shown how fantastically improbable animal-to-man evolution is, evolutionary optimism would not have dominated men's thinking and probably the Marxist would have encountered greater difficulty in making converts.

Linking the idea of human evolution with Naziism, racism, and Marxism does not disprove the idea of human evolution. But if there is such a link, then one is justified in assuming that the evolutionist has contributed, if only by a sin of omission, to the rapid growth of these three ideologies. The Nazi, the racist, and the Marxist have merely carried the evolutionary concept to its logical conclusion. Although they have not depended completely upon Darwin and his successors, neither have they been opposed by that most influential group. If man has indeed evolved, the idea of evolution ought to be carried forward. There is a choice among competing ideologies. Some will wish to accept the idea of a super-race; some will accept the idea of white supremacy; others will choose to follow Marxism, including human evolution in the idea of a perfectible society. There are other possible variations. Most of those who accept evolution seem at present to have a philosophy which is dominated by the idea that man is evolving. This philosophy is not, however, a super-race, white supremacy, or Marxist philosophy. Because of the political activity of these groups, it will be difficult for the evolutionist to continue to reject all three ideologies. The idea of human evolution is explosive, and probably in a few decades this idea will either have died out (an unlikely event) or it will have effected profound and unpleasant changes in our society.

But why suggest that these changes will be unpleasant? One reason is that the changes cited so far have been unpleasant. Another reason is that the idea of human evolution is a structuring principle, and this principle is being imposed on a universe which does not correspond to this structure. Consequently, the imposition will lead to difficulty; the round nail will not fit in the square hole. As the natural man forces on the universe his principle of a self-contained universe, a principle which ultimately forces him to accept the idea of human evolution, he increasingly encounters trouble in science, in politics, and wherever else he applies that principle.

Summary

As with animal-to-animal and plant-to-plant evolution, the scientific evidence for the descent of man from animals is only circumstantial. Theistic evolution is probably a transient position because its one new idea, the idea that God guided in a special way jumps across gaps between kinds, is an unprovable idea, according to the assumptions of theistic evolution itself. The questions about the early history of man should be approached assuming that man did not descend from animals, and that the Biblical data concerning his early history are valid. Accepting the idea of human evolution encourages men to accept evil ideologies, such as Naziism, white supremacy, or Marxism.

REFERENCES

1. Putnam, C., *Race and Reason: a Yankee View,* Public Affairs Press, Washington, D.C., 1961.
2. *Ibid.,* p. 53.

15 Conclusion

Although the natural man and the Christian are diametrically opposed to each other, people tend to be inconsistent. Consequently they sometimes seem to take the position of the natural man, and at other times they seem to take the position of the Christian. Inconsistency in one's world-and-life view is wrong whenever it means that a compromise between these two positions has been made. Christ said to the Laodicean church,

> I know thy works, that thou art neither cold nor hot: I would thou wert cold or hot. So then because thou art lukewarm, and neither cold nor hot, I will spue thee out of my mouth. (Rev. 3:15-16)

The position of the natural man and the Christian are derived from incompatible principles. The root idea of the natural man is that the universe is, in principle, humanly comprehensible, and that the universe is self-contained. The root idea of the Christian is that the Creator-God is outside of his creation, and that the entire Bible is a message from God, who is outside the universe, to men, who are part of the universe.

Summary of the Argument

These two incompatible basic principles lead the thinking of men into opposite directions. If the universe is self-contained and the Bible is no more than a cultural phenomenon, its role in catalyzing the modern explosion of scientific knowledge was a historical accident and not a means whereby God carried out his will in history. The natural man takes man's law, the law in a universe he assumes to be fully comprehensible, to be ultimate. He therefore asserts dogmatically that miracles never occurred because they could not occur. For the Christian, miracles are indications of a law beyond what man can know. By assuming a self-contained universe, the natural man encounters several difficulties. He has no good criterion for accepting evidence and he cannot provide answers to some scientific questions which the Christian can answer, such as the questions concerning the finitude of the universe with respect to space and time. His entire logical structure has collapsed because he finds himself unable to prove the existence of an adequate cause-and-effect law.

Although the Christian is becoming more aware of the unsolvable problems of the natural man, he sometimes depends upon unreliable methods to answer the natural man. With a proper view of creation, the Christian can abandon such procedures. He can realize that God gave him the ability to recognize the hand of God in creation. God made man and the rest of creation in harmony. By means of science man is aided in carrying out the command of God to subdue the earth.

Since the natural man and the Christian are diametrically opposed, it was inevitable that there would be a confrontation concerning origins – the origin of the universe, of the earth, of life, and of man. Although the natural man did not develop all aspects of his position simultaneously he has begun to recognize that all the parts are necessary. Man evolved from animals, he claims; there was animal and plant evolution from the simplest forms of life; life emerged from inanimate matter; and the universe has always existed.

The Christian, with Biblical revelation from God and a more realistic view of scientific evidence, knows that the natural man is wrong on all counts. There has been some over-reaction on the part of some Christians, as they have absolutized certain ideas (such as the recent creation idea) probably not taught by the Bible. Without question the whole evolution question calls for a renewed effort by Christians to evaluate first of all the Biblical evidence, with an accompanying effort to evaluate scientific information from the Christian point of view.

Today the key word in any discussion of belief is "dialogue." If there is an attempt to have dialogue, assuming that the discussants have a common starting point, the dialogue can be successful only if that assumption is true. If the starting points are in fact different, discussion can be meaningful only if *this* difference is recognized. He who denies the Christian position can examine his own position honestly only if he examines his own starting point. An honest appraisal will show him to what extent he has accepted the natural man's basic assumption. Then only can he ask himself which path he will take. He can assume, with the natural man, that nothing exists outside of the physical universe, and that the universe is, in principle, comprehensible. If he is consistent, he will keep this position and follow it to its bitter, frustrating conclusions, as he encounters many dead ends and inconsistencies.

The Christian can *be* a Christian only insofar as he adopts completely the Christian position. He can show the natural man that the correct starting point, the only structuring assumption which corresponds to this universe, is that there is a God outside the universe he has created, and that he has revealed himself in his Bible. As the Christian removes from his thinking that which impedes his understanding of the Bible, he can remove elements of the natural man's thinking from his thoughts. As he allows the Bible to speak to him of creation, of the universe, of man, of God Incarnate, our Lord Jesus Christ, and of whatever else God chooses to teach man through the Bible, the Christian will be brought closer to God. The witness of the Christian to all men will have unbelievable power because his witness will be true.